Role of Agriculture in Implementing Concept of Sustainable Food System

Role of Agriculture in Implementing Concept of Sustainable Food System

Guest Editors

Anna Kocira
Mariola Staniak

MDPI

Basel • Beijing • Wuhan • Barcelona • Belgrade • Novi Sad • Cluj • Manchester

Guest Editors
Anna Kocira
Institute of Human Nutrition and Agriculture
The University College of Applied Sciences in Chełm
Chełm
Poland

Mariola Staniak
Department of Crops and Yield Quality
Institute of Soil Science and Plant Cultivation-State Research Institute
Puławy
Poland

Editorial Office
MDPI AG
Grosspeteranlage 5
4052 Basel, Switzerland

This is a reprint of the Special Issue, published open access by the journal *Agriculture* (ISSN 2077-0472), freely accessible at: https://www.mdpi.com/journal/agriculture/special_issues/Agriculture_Sustainable_Food_System.

For citation purposes, cite each article independently as indicated on the article page online and as indicated below:

Lastname, A.A.; Lastname, B.B. Article Title. *Journal Name* **Year**, *Volume Number*, Page Range.

ISBN 978-3-7258-4189-9 (Hbk)
ISBN 978-3-7258-4190-5 (PDF)
https://doi.org/10.3390/books978-3-7258-4190-5

Contents

About the Editors

Anna Kocira

Anna Kocira has been a researcher at the University College of Applied Sciences in Chełm since 2008. Her research interests include yield-forming, biochemical, bioprotective and the economic effects of biostimulant application. She conducts research on sustainable methods of managing biotic and abiotic stress in plant cultivation. She is particularly interested in the effect of abiotic stresses and cultivation on legume yield and quality, including antioxidant potential and the content of nutrients and fiber. Recognized in the scientific community as a specialist in bean and soybean production, she has served as Head of the Department of Agriculture since 2022. She is the author and co-author of over 100 publications and holds two utility models. She has supervised one doctoral dissertation. She is a member of the Baltic & Black Sea Circle Consortium, the Polish Agronomic Society, and the Chełm Scientific Society. She has also led projects aimed at popularizing science.

Mariola Staniak

Mariola Staniak has been a researcher at the Institute of Soil Science and Plant Cultivation in Puławy since 2003. Her research interests include the biological basis of the productivity of various crop species used for the production of food and animal feed. She is involved in improving agricultural technology, introducing innovations, and adapting treatments to address the challenges related to climate change. She conducts extensive research on the physiological state of crops in response to abiotic stress factors and agrotechnical treatments. She is a recognized specialist in soybean cultivation in Poland and is also interested in issues related to the impact of agriculture on the biodiversity of flora on arable land. From 2016 to 2024, she served as Head of the Department of Fodder Crop Production. She is the author and co-author of over 300 publications, including approx. 150 scientific papers, and has supervised three doctoral dissertations. She is also involved in numerous scientific research projects.

agriculture

MDPI

Editorial

Role of Agriculture in Implementing the Concept of Sustainable Food System

Anna Kocira [1,*,†] and Mariola Staniak [2,*,†]

1 Institute of Human Nutrition and Agriculture, The University College of Applied Sciences in Chełm, 22-100 Chełm, Poland
2 Department of Crops and Yield Quality, Institute of Soil Science and Plant Cultivation—State Research Institute, Czartoryskich 8, 24-100 Puławy, Poland
* Correspondence: akocira@panschelm.edu.pl (A.K.); staniakm@iung.pulawy.pl (M.S.)
† These authors contributed equally to this work.

Nowadays, agriculture faces problems that threaten its basic function, i.e., meeting human needs for food. These problems include: climate change; loss of biodiversity; soil degradation, compaction, salinization and pollution; depletion and pollution of water resources; increasing production costs and decreasing number of farms and rural population [1]. In particular, the intensifying effects of climate change in recent years have contributed to the discussion on the impact of agriculture on these changes [2]. Agriculture is closely linked to the environment, as its efficiency depends on natural resources, and agricultural production using environmental resources can negatively affect its quality [3]. In addition, the impact of agriculture on climate change may result from the fact that it is to some extent responsible for anthropogenic changes in the environment and loss of biodiversity [4]. Therefore, an integrated approach to agricultural production is needed, which includes agriculture that protects land resources (technical and economic) and is friendly to the natural environment (agroecology-rural), which has a positive impact on the protection of biodiversity and contributes to more sustainable agriculture [5]. The effect of these activities is the promotion of more sustainable practices from food production to consumption, leading to the implementation of innovative solutions in agricultural practice and directly affecting the quality of food.

The changes occurring in the natural environment oblige agricultural production to be conducted in accordance with the principles of sustainable development. Recently, this has become a priority for industrialized countries, where the development of the agricultural sector previously proceeded in accordance with the principles of industrial agriculture [6]. Changes in agriculture towards sustainable development should take place by promoting innovative technologies and management models [7]. It has been confirmed that the impact of agriculture on the natural environment depends on the intensity of its production. Therefore, it is necessary to promote the intensification of agriculture in accordance with the principles of sustainability, the so-called ecological intensification [8].

Sustainable agriculture requires the introduction of practices that support the protection and preservation of biodiversity and the ecosystem services resulting from it, improving soil health and biodiversity, reducing greenhouse gas emissions, increasing carbon sequestration in the soil by using diversified crop rotations, cover crops, creating protective belts for wildlife or using agroecological cultivation methods [9].

In the context of implementing the concept of sustainable development, agricultural production focused on production systems that are safe for people and the environment, which includes organic farming, is important. Organic farming, compared to conventional farming, is perceived as more environmentally friendly due to the reduction of carbon and ammonia emissions, positive impact on soil conditions (increased organic matter content and biological activity of the soil, reduced soil erosion) and water conditions (reduced leaching of nitrates and pesticides into groundwater and surface water) [10,11].

The implementation of these practices is important both at the central and local levels, which aims to reduce the negative effects of environmental pollution and the ongoing global warming. The level of development of organic production is different among the members of the European Union, and the main differences are related to the number of farms and the area of organic land. Therefore, the need for a non-uniform approach to sustainable agriculture in the EU, with particular emphasis on the differences between the old and new Member States, is emphasized by Zakrzewska et al. [12]. The challenges related to changes in the development of agriculture should be different depending on the level of productivity and the amount of inputs in individual Member States. However, the main challenge should be to strive for a balance between economic, social and environmental goals in agricultural production.

Long-term use of sustainable agricultural practices brings measurable effects. Particularly important are practices that improve the health and biodiversity of agroecosystems, increase carbon sequestration in the soil, or reduce greenhouse gas emissions, which include the use of diversified crop rotations, including the introduction of legumes to the crop rotation, cover crops, minimum tillage or integration of grazing livestock into crop production systems [13–15].

A significant problem in plant cultivation is also abiotic stress factors, which significantly reduce the quantity and quality of the crop. Therefore, the application of natural biopreparations (containing free amino acids, humic compounds, seaweed or plant extracts, chitin, chitosan, or microbial inoculants) that have a beneficial effect on plants by stimulating plant growth, improving mineral uptake and increasing plant tolerance to biotic and abiotic stresses, is one of the methods supporting plant production, also in the ecological system [16,17]. It is important to use these biopreparations in plant cultivation intended for both food and animal feed, which contributes to improving the quality of agricultural crops.

In turn, in animal production, the basis for sustainable breeding is to increase productivity while reducing negative effects on the environment and improving the welfare of farm animals [18,19]. One of the elements of sustainable production is to reduce CO_2 emissions, and one of the strategies to introduce these changes is the introduction of breeding indices with trait weights derived from the agricultural environment and genetic selection of individual animals based on economic and carbon indicators [20,21].

Modern agriculture faces the need to reconcile the production of high-quality food with care for the environment. In this context, one of the key challenges is the effective and responsible management of agrophages. Traditional approaches to pest manage-ment, mainly based on the use of pesticides, have contributed to increased yields in re-cent decades, but at the expense of negative environmental impacts, such as soil and water pollution, decline in biodiversity, development of agrophage resistance, and health effects [22]. Sustainable food production systems integrate biological, agrotechnical, physical and chemical crop protection methods in a way, that minimises risks to the environment and human and animal health, but at the same time takes into ac-count social and economic aspects [23]. In modern agriculture, advanced agricultural technologies and communication tools also play an important role in food production [24]. Modern monitoring systems for pests can be important decision-support tools for the timing and need for protective treatments. Continuous, automated monitoring can provide a wealth of useful information, including species composition, timing of pest emergence and seasonal activity. Integrating phenological data with meteorological data allows for more precise planning of control treatments, thus reducing the use of chemicals and their pressure on the environment.

The authors also highlight the role of systematic monitoring using various tools such as pheromone traps and sticky tapes in determining the appropriate timing of pesticide application. They point out, that the efficacy of an agent is highly dependent on the developmental stage of the pest at which it was applied, for which a good knowledge of the pest's biology and behavior is essential. The authors point out the great importance of biological methods of protection in preventing the development of pest resistance to some chemical plant protection products. The use of biological methods is also a response to environmental requirements and new regulations related to the objectives of the European Green Deal and the Farm to fork strategy.

An interesting study on the potential use of some plant species in the phytoremediation of degraded agrocenoses is presented by Szmagara et al. [25]. The aim of the study was to investigate the resistance of ten varieties of canna (*Canna indica*) to fungal diseases, but also their ability to grow in contaminated environments. The results con-firmed the high tolerance of some varieties to environmental stresses and infestation by fungal pathogens (mainly *Fusarium* spp., *Alternaria alternata*), while maintaining favorable photosynthesis and gas exchange parameters. This may indicate the potential use of such species on degenerated soils, in regenerative agriculture or in marginal land management. In addition, the resistance of some varieties to fungal pathogens gives them the additional function of reducing the amount of chemical protection used and thus reducing the pressure on the environment.

Conclusions from the research discussed in this Special Issue point to the need for further development of early warning systems, monitoring of agrophages and in-tegration of environmental data into cultivation practices. Sustainable agrophage management requires a systematic approach, combining knowledge of species biology, weather data, digital technologies and elements of biological control to increase con-servation efficiency and reduce the negative environmental impact of agriculture.

The concept of sustainability involves minimising the negative impact of agriculture on the environment while ensuring food security and the economic viability of farms [26]. In this context, modern technical solutions play an important role which, on the one hand, increase production efficiency and, on the other, can reduce energy consumption, material consumption or greenhouse gas emissions [2]. Among the innovations in agriculture, we can mention precision crop management systems, modern materials, machinery or technological processes. The development of remote monitoring tools, field sensors and predictive models enables rational management of water, fertilizers and plant protection products, yield fore-casting or early detection of diseases and pests [27,28]. This is supported by a study by Kulig et al. [29], in which the possibility of using remote sensing methods based on UAV and hand-held devices was assessed for yield forecasting of winter wheat with reduced allergenic proteins. Using modern equipment, plant condition was monitored to optimize fertilization without yield loss. The study showed differences in the response of individual genotypes (with and without allergenic protein) to fertilization levels. Statistical analysis showed, that indices based on radiometric measurements have a moderate correlation with grain yield, but differences in prediction are not significant. The authors point to the need to continue field studies dedicated to monitoring and evaluating grain productivity, especially using UAV remote sensing techniques.

Also of great importance are new generations of agricultural materials, e.g., biodegradable food packaging, which make food production more sustainable [30,31]. In this context, the research on the evaluation of a new generation of agricultural nets (Tama LT) for wrapping various bulk materials made of light and strong HDPE polymers is interesting. The study showed, that nets produced with 'light' technology provide high quality wrapping of e.g., hay, straw, green fodder, and their use may contribute to reducing the amount of plastics used in agriculture. The authors mention the urgent need to develop and implement technologies for recovering used agricultural nets and converting them into granulate, that can be reused in production.

Both the use of remote crop monitoring technology and the introduction of modern agricultural materials show, that the technical aspects of agricultural production can significantly support sustainability goals. The introduction of modern agricultural solutions enables more precise crop management, which translates into efficient and less environmentally damaging food production.

Sustainable agriculture allows for the implementation of environmental and climate protection goals while maintaining food production at the current level and, in the long term, even increasing it. The articles presented in this reprint comprehensively cover issues related to the introduction of sustainable agricultural practices. The proposals for changes in technologies presented in them will allow for agricultural production in accordance with the principles of sustainable development. These articles concern both plant and animal production, with particular emphasis on ecological aspects while maintaining efficiency. It has been shown that it is justified to promote long-term practices based on responsible management of natural resources promoting sustainable development.

Author Contributions: A.K. and M.S. contributed equally to this article. All authors have read and agreed to the published version of the manuscript.

Acknowledgments: The Guest Editors wish to thank all the Authors for their contribution to this Special Issue. We also want to thank the Reviewers, Editorial Managers and Editors who assisted in developing this Special Issue.

Conflicts of Interest: The authors declare no conflicts of interest.

References

1. Velten, S.; Leventon, J.; Jager, N.; Newig, J. What is sustainable agriculture? A systematic review. *Sustainability* **2015**, *7*, 7833–7865. [CrossRef]
2. Faber, A.; Jarosz, Z. Impact of climate change on agriculture in Poland. *Curr. Agron.* **2024**, *53*, 1–9. [CrossRef]
3. Wang, G.; Shi, R.; Mi, L.; Hu, J. Agricultural Eco-Efficiency: Challenges and Progress. *Sustainability* **2022**, *14*, 1051. [CrossRef]
4. Ortiz-Bobea, A.; Ault, T.R.; Carrillo, C.M.; Chambers, R.G.; Lobell, D.B. Anthropogenic climate change has slowed global agricultural productivity growth. *Nat. Clim. Change* **2021**, *11*, 306–312. [CrossRef]
5. Fischer, J.; Brosi, B.; Daily, G.C.; Ehrlich, P.R.; Goldman, R.; Goldstein, J.; Lindenmayer, D.B.; Manning, A.D.; Mooney, H.A.; Pejchar, L.; et al. Should agricultural policies encourage land sparing or wildlife-friendly farming? *Front. Ecol. Environ.* **2008**, *6*, 380–385. [CrossRef]
6. Czyżewski, A.; Staniszewski, J. Dilemmas of operationalising the paradigm of sustainable agricultural development using the concept of eco-efficiency. *Probl. World Agric.* **2018**, *18*, 44–56. (In Polish) [CrossRef]
7. Nowak, A.; Krukowski, A.; Różańska-Boczula, M. Assessment of sustainability in agriculture of the European Union countries. *Agronomy* **2019**, *9*, 890. [CrossRef]
8. Kleijn, D.; Bommarco, R.; Fijen, T.P.; Garibaldi, L.A.; Potts, S.G.; Van Der Putten, W.H. Ecological intensification: Bridging the gap between science and practice. *Trends Ecol. Evol.* **2019**, *34*, 154–166. [CrossRef]
9. Żuchowska-Grzywacz, M. 2024 Sustainable agriculture in the face of contemporary challenges. *Zesz. Prawnicze* **2024**, *24*, 123–143. (In Polish) [CrossRef]
10. Meemken, E.; Qaim, M. Organic Agriculture, Food Security, and the Environment. *Annu. Rev. Resour. Econ.* **2018**, *10*, 39–63. [CrossRef]
11. Lorenz, K.; Lal, R. Environmental Impact of Organic Agriculture. *Adv. Agron.* **2016**, *139*, 99–152. [CrossRef]
12. Zakrzewska, A.; Nowak, A. Diversification of Agricultural Output Intensity across the European Union in Light of the Assumptions of Sustainable Development. *Agriculture* **2022**, *12*, 1370. [CrossRef]
13. Sharma, P.; Singh, A.; Kahlon, C.S.; Brar, A.S.; Grover, K.K.; Dia, M.; Steiner, R.L. The role of cover crops towards sustainable soil health and agriculture—A review paper. *Am. J. Plant Sci.* **2018**, *9*, 1935–1951. [CrossRef]

14. Shah, K.K.; Modi, B.; Pandey, H.P.; Subedi, A.; Aryal, G.; Pandey, M.; Shrestha, J. Diversified crop rotation: An approach for sustainable agriculture production. *Adv. Agric.* **2021**, *2021*, 8924087. [CrossRef]
15. Rempelos, L.; Baranski, M.; Wang, J.; Adams, T.N.; Adebusuyi, K.; Beckman, J.J.; Brockbank, C.J.; Douglas, B.S.; Feng, T.; Greenway, J.D.; et al. Integrated soil and crop management in organic agriculture: A logical framework to ensure food quality and human health? *Agronomy* **2021**, *11*, 2494. [CrossRef]
16. Krawczuk, A.; Ogrodniczek, J.; Bohata, A.; Bartos, P.; Olšan, P.; Findura, P.; Kocira, S. Physical Properties of Plant Extracts with Biostimulant Potential Produced Using Cold Plasma and Low-Pressure Microwave Discharge. *Agric. Eng.* **2024**, *28*, 277–285. [CrossRef]
17. Kocira, S.; Bohatá, A.; Bartoš, P.; Olšan, P.; Pérez-Pizá, M.C.; Świeca, M.; Sozoniuk, M.; Szparaga, A.; Bedrníček, J.; Lorenc, F.; et al. Technologies for producing Plant Biostimulants using Cold Plasma and Low-Pressure Microwave Discharge. *Agric. Eng.* **2024**, *28*, 341–351. [CrossRef]
18. Clay, N.; Garnett, T.; Lorimer, J. Dairy intensification: Drivers, impacts and alternatives. *Ambio* **2020**, *49*, 35–48. [CrossRef]
19. Brito, L.F.; Bedere, N.; Douhard, F.; Oliveira, H.R.; Arnal, M.; Peñagaricano, F.; Schinckel, A.P.; Baes, C.F.; Miglior, F. Review: Genetic selection of high-yielding dairy cattle toward sustainable farming systems in a rapidly changing world. *Animal* **2021**, *15*, 100292. [CrossRef]
20. Richardson, C.M.; Sunduimijid, B.; Amer, P.; van den Berg, I.; Pryce, J.E. A method for implementing methane breeding values in Australian dairy cattle. *Anim. Prod. Sci.* **2021**, *61*, 1781–1787. [CrossRef]
21. Kelleher, M.M.; Amer, P.R.; Shalloo, L.; Evans, R.D.; Byrne, T.J.; Buckley, F.; Berry, D.P. Development of an index to rank dairy females on expected lifetime profit. *J. Dairy Sci.* **2015**, *98*, 4225–4239. [CrossRef]
22. Abrol, D.P.; Shankar, U. Persticides, Food Safety and Integrated Pest Management. In *Integrated Pest Mangement-Pesticide Problems*; Pimentel, D., Peshin, R., Eds.; Springer: New York, NY, USA, 2014; Volume 3, pp. 167–199.
23. Naranjo, S.; Ellsworth, P. Fifty years of the integrated control concept: Moving the model and implementation forward in Arizona. *Pest Manag. Sci.* **2009**, *65*, 1267–1286. [CrossRef] [PubMed]
24. Dara, S.K. The new integrated pest management paradigm for the modern age. *J. Integr. Pest Manag.* **2019**, *10*, 1–9. [CrossRef]
25. Szmagara, M.; Kopacki, M.; Skwaryło-Bednarz, B.; Jamiołkowska, A.; Marcinek, B.; Rysiak, K.; Szmagara, A. Assessment of Biometric Parameters and Health of Canna's Cultivars as Plant Useful in Phytoremediation of Degraded Agrocenoses. *Agriculture* **2023**, *13*, 157. [CrossRef]
26. Berbeć, A.K. Agricultural resilience and agricultural sustainability—Which is which? *Curr. Agron.* **2024**, *53*, 10–22. [CrossRef]
27. Sishodia, R.P.; Ray, R.L.; Singh, S.K. Applications of Remote Sensing in Precision Agriculture: A Review. *Remote Sens.* **2020**, *12*, 3136. [CrossRef]
28. Huang, Y.; Chen, Z.-X.; Yu, T.; Huang, X.-Z.; Gu, X.-F. Agricultural remote sensing big date: Management and applications. *J. Integr. Agric.* **2018**, *17*, 1915–1931. [CrossRef]
29. Kulig, B.; Waga, J.; Oleksy, A.; Rapacz, M.; Kołodziejczyk, M.; Wężyk, P.; Klimek-Kopyra, A.; Witkowicz, R.; Skoczowski, A.; Podolska, G.; et al. Forecasting of Hypoallergenic Wheat Productivity Based on Unmanned Aerial Vehicles Remote Sensing Approach—Case Study. *Agriculture* **2023**, *13*, 282. [CrossRef]
30. Alaswad, A.O.; Mahmoud, A.S.; Arunachalam, P. Recent Advances in Biodegradable Polymers and Their Biological Applications: A Brief Review. *Polymers* **2022**, *14*, 4924. [CrossRef]
31. Siracusa, V.; Rocculi, P.; Romani, S.; Rosa, M.D. Biodegradable Polymers for Food Packaging: A Review. *Trends Food Sci. Technol.* **2008**, *19*, 634–643. [CrossRef]

agriculture

MDPI

Article

Effect of Humus, Compost, and Vermicompost Extracts on the Net Energy Concentration, Net Energy of Lactation, and Energy Yield of *Dactylis glomerata* and *Lolium perenne*

Jacek Sosnowski, Milena Truba * and Katarzyna Jarecka

Institute of Agriculture and Horticulture, Siedlce University of Natural Sciences and Humanities, Prusa 14, 08-110 Siedlce, Poland; jacek.sosnowski@uph.edu.pl (J.S.); kj33@stud.uph.edu.pl (K.J.)
* Correspondence: milena.truba@uph.edu.pl

Abstract: The purpose of the research was to examine whether selected organic materials could increase the net energy concentration, net energy of lactation, and energy yield of two grass species. The main factors in the experiment were soil conditioners with the content based on compost, vermicompost and humus extract, applied separately and supplemented with NPK fertilizers. The second factor is grass species, *Dactylis glomerata* and *Lolium perenne*. Mineral fertilizer and compost extract resulted in a reduction in net energy concentration by about 5%. The largest value of net energy of lactation (NEL) was obtained after the application of humus extract with NPK. The average value of the energy yield was the biggest on units where biological extracts were applied together with NPK. That increase was from 63% for NPK applied together with humus extract to 76.5% for the NPK applied with compost extract. Of the applied humic substances, those applied together with mineral fertilizer had the greatest impact on net energy value and NEL. The use of compost extract contributed to a substantial increase in the yield of feed energy. Other biological substances applied together with mineral fertilizer also had a positive impact. Weather conditions in various years of the research differentiated feed energy values. Due to the complexity of the environment (soil, climate, plant), it is impossible to provide a universal combination of fertilizers that increases the energy value of forage. Therefore, it is important to carry out practical field experiments that will indicate the optimal fertilizer combinations suitable for the selected region.

Keywords: grass; feed; NEL; DLG standard

1. Introduction

Systems for feeding dairy cows (DLG) intended to balance energy and nutrients for high performance cows have been applied in Poland since 1999. The use of the standards to formulate basic rations for livestock brings benefits, increasing productivity and reducing production costs. Based on net energy of lactation (NEL) in the case of dairy cows, the DLG system, and a variety of others, determines the value of the energy of feed stuffs [1]. According to Jonker et al. [2], the content of energy and protein in the correctly balanced ration affects an optimal course of lactation and milk composition. Juszczak and Ziemiński [3] reported that animals at the height of the lactation period suffer from low energy and protein content in the feed when nutritional needs outweigh the possibility of nutrient intake to maintain nutritional balance. However, an excess of protein in the feed can also be harmful as it can cause an increase in the number of milk somatic cells. As the literature indicates [4,5], feed energy value varies depending on the plant species, growing season, fertilizer used, irrigation, and the stage at which the plants are harvested.

Cocksfoot grass is successfully cultivated in unfavourable hydrothermal conditions [6,7]. The literature reports that cocksfoot grows well in drought conditions [8] and under water stress [9]. Research conducted on acidic soils of Lithuania showed that the highest yield of cocksfoot was obtained with high nitrogen fertilization with simultaneous liming [10]. In turn, perennial ryegrass is sensitive to high soil moisture and drought stress [11]. The yield of ryegrass increases with the phosphorus content in the soil; therefore, this species is used for the remediation of soils with a high content of P [12]. Nevertheless, the soil factor was omitted from the experiment, which focused instead on the effects of fertilization and the weather conditions.

In the era of organic farming, which excludes the use of mineral fertilizer, biological growth-enhancing products are increasingly used. Another reason to use of biological preparations is the reduction in mineral fertilization by 20% in the EU by 2030 [13]. In Poland, biological products are enumerated in the list of fertilizers and soil conditioners drawn up by the Institute of Soil Science and Plant Cultivation (IUNG) in Puławy. Although there have been some research studies concerning the effect of soil amendments on the quantitative and qualitative properties of grass [14–19], there is no information available in the literature on the application of biological products to improve the energy value of grass.

Three biological preparations with different compositions were selected for the study: compost extract, vermicompost extract and humus extract. In order to verify the validity of the use of biopreparations, they were compared with NPK mineral fertilization. Additionally, biological preparations were combined with mineral fertilization to see if their effect would be better together or separately. The aim of the research was to determine how biological preparations used separately and in combination with NPK affect the net energy concentration, net energy lactation and the energy yield of two grass species.

2. Materials and Methods

2.1. Experiment Location

The experiment was set up in the autumn of 2018. The three-year research was conducted in the experimental field at the University of Natural Sciences and Humanities in Siedlce (52.169° N, 22.280° E), Poland. With a split-plot arrangement, the experimental units were plots of 3 m^3 with three replications.

The study was conducted on the soil with the granulometric composition of light loamy sand, classified as technosol [20]. Chemical analysis showed that it was of slight acidic pH (pH = 6.6), with a concentration of C_{org} of 12.30 g kg^{-1} DM and N_{total} of 1.250 g kg^{-1} DM. The assimilable macronutrients concentration (mg kg^{-1} DM) was P–790; K–1060; Mg–1260; Ca–1820.

2.2. Experimental Factors

The main factors in the experiment were biological fertilizers, with the trade names of UGmax (compost extract, CE), Eko-Użyźniacz (vermicompost extract, VE), and Humus Active Papka (humus extract, HE), applied separately and supplemented with NPK fertilizers, according to the Institute of Soil Science and Plant Cultivation in Puławy, and their composition is presented in Table 1. In the present experiment, they were tested on two fodder grass species, *Dactylis glomerata* var. Bora and *Lolium perenne* var. Info, sown in the autumn of 2018 with sowing rates of 18 and 23 kg ha^{-1}, respectively. They were used each year in the spring before the growing season with the following doses: compost extract-0.6 dm^3 ha^{-1}, vermicompost extract-15 dm^3 ha^{-1} and humus extract-50 dm^3 ha^{-1}.

Mineral nitrogen, phosphorus, and potassium (NPK) fertilizers were used at the following doses: N–150, P (P_2O_5)–80, K (K_2O)–120 kg ha^{-1}.

Table 1. Soil conditioner composition based on manufacturers' data.

Name	CE	VE	HE
	Macronutrients (g kg^{-1})		
N	1.2	0.6	0.2
P	0.2	0.3	1.3
K	2.9	0.7	4.6
Ca	-	-	3.0
Mg	0.1	-	0.5
Na	0.2	-	-
	Micronutrients (mg kg^{-1})		
Mn	0.3	-	15
Fe	-	-	500
Zn	-	-	3
Cu	-	-	1
Mo	-	-	-
	Microorganisms		
	lactic acid bacteria, photosynthetic bacteria, *Azotobacter, Pseudomonas*, yeast, *Actinomycetes*	Endo micorrhiza, fungi, bacteria, enzymes of earthworms	Useful microorganisms

2.3. Weather Conditions

Meteorological data for the years of research were obtained from the Hydrological and Meteorological Station in Siedlce (Table 2).

Table 2. Average air temperature and sum of atmospheric precipitation in consecutive months of the growing seasons.

Year	Month							
	Apr.	May	June	July	Aug.	Sept.	Oct.	Means
			Temperature (°C)					
2019	13.1	17.0	18.3	20.4	20.6	15.9	9.6	16.4
2020	8.6	11.7	19.3	19.0	20.2	15.5	12.0	15.2
2021	6.6	12.4	20.4	22.7	17.1	12.9	8.6	14.4
Means	9.4	13.7	19.3	20.7	19.3	14.8	10.1	15.3
Multiannual means	8.5	14.0	17.4	19.8	18.9	13.2	7.9	14.2
			Precipitation (mm)					
2019	5.9	59.8	35.9	29.7	49.3	17.4	9.5	29.6
2020	6.0	63.5	118.5	67.7	18.0	38.8	17.6	47.2
2021	42	30	34	50	95	42	6	42.7
Means	18.0	51.1	62.8	49.1	54.1	32.7	11.0	39.8
Multiannual means	33.0	52.0	52.0	65.0	56.0	48.0	28.0	47.7

In the first year (2019), optimal precipitation was only in May and August. In the remaining months of that growing season, rainfall was at least twice as low as the annual mean. The years 2020 and 2021 were rich in rainfall, but dry periods also occurred. The average temperature in 2019 was about 13% higher than the average temperature according to the annual mean. The temperatures recorded in 2020 and 2021 were close to the annual average.

2.4. Analysis

Net energy concentration in 1 kg of dry matter was determined using the following formula [21]:

$$NE = 1.50 - 0.02 \cdot CF \quad (1)$$

where:

NE—Net energy concentration in 1 kg DM,

CF—Crude fibre content (% DM).

Net energy of lactation was determined with the following formula [22]:

$$NEL = 6.998 - 0.061 \cdot CF + 0.014 \cdot TP \quad (2)$$

where:

NEL—net energy of lactation (MJ kg^{-1} DM),
CF—crude fibre content (% DM),
TP—total protein (% DM).

Energy yield of the fodder was determined using the following formula [5]:

$$PE = P \cdot 100\ (0.968 - 0.0063 \cdot CF + 0.033 \cdot TP) \quad (3)$$

where:

PE—forage energy yield (JP ha^{-1}),
P—dry matter yield (dt ha^{-1}),
CF—crude fibre content (% DM),
TP—total protein content (% DM).

During each of the three growth cycles, the plants were cut three times per year (May, July and September). During plant harvest, the green mass of each plot was cut and weighed. Then, samples of the plant material (1.0 kg on average) were taken for chemical analyses. The dry weight of plants was determined by the drying and weighing method. For chemical analyses, the dry plant raw material was ground (including leaves, stems and inflorescences).

The content of total protein and crude fibre in plant material was measured with near-infrared spectroscopy, using the NIRFlex N-500 spectrometer (BUCHI, Flawil, Switzerland) with the INGOT calibration package for dry feed.

The results of the research were processed statistically using three factor analysis of variance. The significance of the impact of experimental factors on the tested characteristics was verified with the Fisher–Snedecor test, while Tukey's test was used to evaluate differences between means. The calculations were conducted with the Statistica 13 Program (TIBCO Software Inc., Palo Alto, CA, USA).

3. Results and Discussion

3.1. Net Energy Concentration in 1 kg DM

Both grass species differed in their average concentration of net energy (NE) in 1 kg of dry matter; it was significantly greater by 7.5% in *Lolium perenne* (1.098) than in *Dactylis glomerata* (1.022). The latter did not respond to different fertilizer treatments in a statistically significant way (Table 3). In turn, *Lolium perenne* had the largest NE concentration on the plots where vermicompost extract was applied (1.145) and where humus concentrate was used together with mineral fertilizer (1.121). Mineral fertilizer and compost extract, both applied separately, resulted in a reduction in this parameter by about 5% compared to the control; it was not, however, a statistically significant difference. Analysing the response to all treatments, it was found that the largest concentration of NE as an average for both species was after vermicompost application (1.082), and after treatment with a combination of humus extract and mineral fertilizer (1.084). According to Wiśniewska-Kadżajan [21], applying manure and mushroom substrate both alone and with mineral fertilization, it was found that the net energy concentration in forage ranged from 0.93 to 0.95, and different kinds of treatment did not differentiate the values significantly. In the present experiment, the highest increase in NE concentration (1.076) was observed in the second growing season in 2020 (Table 3). The abundance of rainfall and moderate temperatures in 2020 could have contributed to the accumulation of NE in the plants. This was supported by a decrease of 4% in NE in seasons (years) when dry periods and higher temperatures prevailed.

Table 3. Net energy concentration in 1 kg DM.

	Fertiliser Effect								Means
	0	NPK	CE	VE	HE	CE + NPK	VE + NPK	HE + NPK	
	Means for species								
Dactylis glomerata	1.010 ^{Ba}	1.020 ^{Aa}	0.993 ^{Ba}	1.018 ^{Ba}	1.037 ^{Ba}	1.026 ^{Ba}	1.022 ^{Ba}	1.047 ^{Ba}	1.022 ^{B}
Lolium perenne	1.104 ^{Aab}	1.048 ^{Ab}	1.053 ^{Ab}	1.145 ^{Aa}	1.113 ^{Aab}	1.109 ^{Aab}	1.089 ^{Aab}	1.121 ^{Aa}	1.098 ^{A}
	Means for growing seasons								
2019	1.030 ^{Aa}	1.032 ^{Aa}	1.045 ^{Aa}	1.097 ^{Aa}	1.080 ^{Aa}	1.094 ^{Aa}	1.053 ^{Aa}	1.088 ^{Aa}	1.065 ^{AB}
2020	1.105 ^{Aa}	1.072 ^{Aa}	1.037 ^{Aa}	1.083 ^{Aa}	1.063 ^{Aa}	1.050 ^{Aa}	1.083 ^{Aa}	1.116 ^{Aa}	1.076 ^{A}
2021	1.037 ^{Aa}	0.997 ^{Aa}	0.987 ^{Aa}	1.066 ^{Aa}	1.082 ^{Aa}	1.058 ^{Aa}	1.030 ^{Aa}	1.048 ^{Aa}	1.038 ^{B}
	Means for harvests								
I	1.035 ^{Aa}	0.971 ^{Ba}	1.005 ^{Ba}	1.054 ^{Aa}	0.997 ^{Ba}	1.033 ^{Aa}	1.014 ^{Aa}	1.055 ^{Aa}	1.020 ^{B}
II	1.035 ^{Aa}	1.038 ^{ABa}	1.091 ^{ABa}	1.050 ^{Aa}	1.041 ^{ABa}	1.077 ^{Aa}	1.076 ^{Aa}	1.082 ^{Aa}	1.061 ^{AB}
III	1.102 ^{Aa}	1.100 ^{Aa}	1.112 ^{Aa}	1.094 ^{Aa}	1.089 ^{Aa}	1.091 ^{Aa}	1.077 ^{Aa}	1.114 ^{Aa}	1.097 ^{A}
Mean	1.057 ^{ab}	1.034 ^{ab}	1.023 ^{b}	1.082 ^{a}	1.075 ^{ab}	1.067 ^{ab}	1.056 ^{ab}	1.084 ^{a}	

0—Control; NPK—mineral fertiliser; CE—compost extract; VE—vermicompost extract; HE—humus extract; Means in lines marked with the same small letters do not differ significantly; Means in columns marked with the same capital letters do not differ significantly.

The concentration of energy in 1 kg DM of plants depends, to a large extent, on weather conditions. Grass species displayed different sensitivity to changing weather during the first growing season (Figure 1a) when *Dactylis glomerata* had the lowest concentration of NE (1.007), while *Lolium perenne* had the highest (1.123). NE concentration in both species decreased in the last year (2021), which could have been caused by alternately occurring dry and wet periods. NE was the smallest in the first harvest (1.020) and then increased with successive ones by about 3.5% to its maximum in plants of the third cut (1.097). For both grass species, net energy concentration also increased in subsequent harvests, and the difference between the first and the last was about 7%, being statistically significant (Figure 1b).

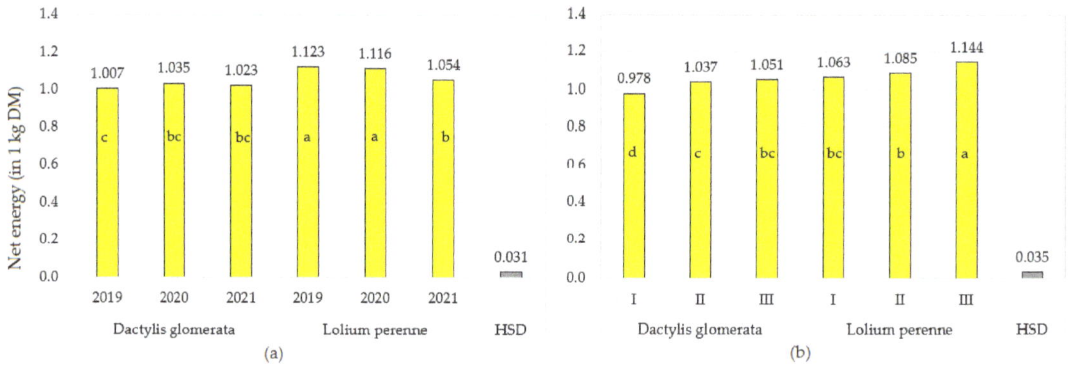

Figure 1. Net energy of *Dactylis glomerata* and *Lolium perenne* in consecutive (**a**) growing seasons and (**b**) harvests (in 1 kg DM). Means in columns marked with the same lower-case letters do not differ significantly.

3.2. Net Energy of Lactation

The average net energy of lactation (NEL) was greater in *Lolium perenne* forage (5.98 MJ kg^{-1} DM), and the 4% difference between both grass species was statistically significant (Table 4). The NEL results for the two species ranged from 5.64 to 6.12 MJ kg^{-1}, which classified them as good quality forage [1]. According to Abas et al. [4], NEL for grass hay was 3.78 MJ kg^{-1} and for alfalfa hay 5.20 MJ kg^{-1}. Analysing the response of the species to different treatments, the value of the NEL of *Dactylis glomerata* showed no significant variation. In turn, *Lolium perenne* had the largest NEL on units where vermicompost extract was applied (6.12 MJ kg^{-1} DM). The responses were similar in the case of humic extract (6.03 MJ kg^{-1} DM), compost extract in combination with mineral fertilization (6.03 MJ kg^{-1} DM), and humic extract applied with mineral fertilizer (6.06 MJ kg^{-1} DM). According to Kujawiak and Zarudzki [1], forage with the NEL value from 6.0 to 6.5 MJ kg^{-1} DM is of very good quality.

Table 4. Net energy of lactation of *Dactylis glomerata* and *Lolium perenne* in consecutive harvests and growing seasons (MJ·kg^{-1} DM).

	Fertiliser Effect								Mean
	0	NPK	CE	VE	HE	CE + NPK	VE + NPK	HE + NPK	
	Means within species								
Dactylis glomerata	5.70 Ba	5.73 Aa	5.64 Aa	5.73 Ba	5.79 Ba	5.77 Ba	5.76 Ba	5.83 Ba	5.74 B
Lolium perenne	5.99 Aab	5.82 Ab	5.80 Ab	6.12 Aa	6.03 Aa	6.03 Aa	5.97 Aab	6.06 Aa	5.98 A
	Means within growing seasons								
2019	5.74 Aa	5.77 Aa	5.77 Aa	5.96 Aa	5.91 Aa	5.98 Aa	5.86 Aa	5.96 Aa	5.87 AB
2020	5.99 Aa	5.90 Aa	5.78 Aa	5.93 Aa	5.88 Aa	5.83 Aa	5.96 Aa	6.05 Aa	5.92 A
2021	5.80 Aa	5.65 Aa	5.62 Aa	5.88 Aa	5.94 Aa	5.88 Aa	5.77 Aa	5.83 Aa	5.80 B
	Mean within harvests								
I	5.76 Aa	5.56 Ba	5.64 Ba	5.82 Aa	5.65 Ba	5.78 Aa	5.72 Aa	5.85 Aa	5.72 B
II	5.76 Aa	5.78 Aa	5.93 Aa	5.82 Aa	5.81 Aa	5.93 Aa	5.93 Aa	5.94 Aa	5.86 AB
III	6.01 Aa	6.01 Aa	6.01 Aa	5.98 Aa	5.97 Aa	5.98 Aa	5.94 Aa	6.06 Aa	6.00 A
Mean	5.84 ab	5.77 b	5.72 b	5.92 ab	5.91 ab	5.90 ab	5.86 ab	5.95 a	

0—Control; NPK—mineral fertiliser; CE—compost extract; VE—vermicompost extract; HE—humus extract; Means in lines marked with the same small letters do not differ significantly; Means in columns marked with the same capital letters do not differ significantly.

By comparing the values of NEL for different treatments, as an average for both species, the largest was obtained after the application of humus extract with mineral fertilizer (5.95 MJ kg^{-1} DM). Mineral fertilizer applied on its own did not increase it (5.77 MJ kg^{-1} DM), and neither did compost extract (5.72 MJ kg^{-1} DM). Those values do not differ significantly in terms of statistical significance. However, the average NEL values for fertilization indicated that the feed was of good quality [1]. On average, the largest NEL was observed in the second year (5.92 MJ kg^{-1} DM), with a significant reduction in the third year (5.80 MJ kg^{-1} DM). The differences in the net energy of lactation content between different growing seasons were probably caused by weather conditions. The results of the research indicated that wet periods during a growing season promoted the accumulation of net energy of lactation in grass forage, while dynamic changes in meteorological conditions, as in 2021, decreased it.

NEL in different growing seasons varied depending on the grass species (Figure 2a). In *Dactylis glomerata* fodder, this parameter remained at a similar level (5.69–5.78 MJ ha^{-1} DM), not showing significant differences in all three growing seasons. In turn, for the feed of *Lolium perenne,* the largest NEL value was recorded in the first (6.06 MJ ha^{-1} DM) and second (6.05 MJ ha^{-1} DM) years of the research, while in the third this parameter decreased considerably to 5.83 MJ ha^{-1} DM, i.e., by about 3.6%. The greatest value of net energy of lactation was in the last harvest (5.84 and 6.15 MJ ha^{-1} DM), and the smallest in the first (5.59 and 5.85 MJ ha^{-1} DM). Both grass species had the same tendency of increasing the value of the NEL parameter from the first to third harvest by about 4.5% on average (Figure 2b).

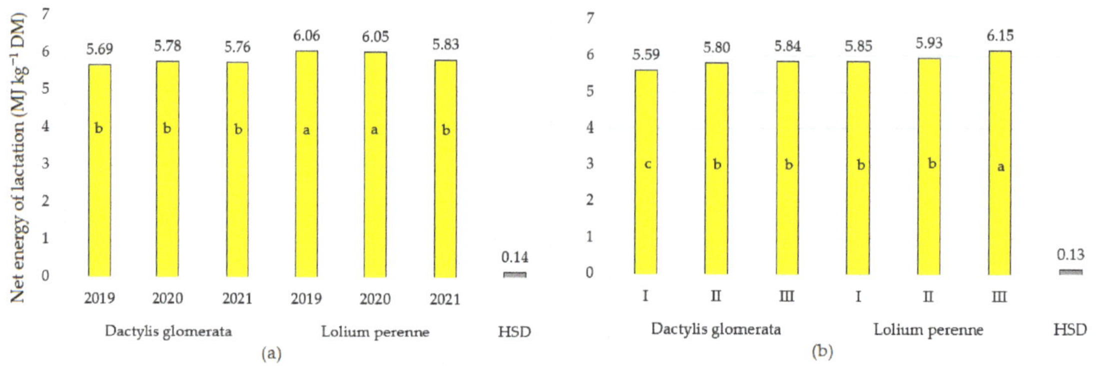

Figure 2. Net energy of lactation of *Dactylis glomerata* and *Lolium perenne* in consecutive (**a**) growing seasons (**b**) harvests (MJ kg^{-1} DM). Means in columns marked with the same lower-case letters do not differ significantly.

3.3. The Yield of Feed Energy

Feed energy yields vary depending on the plant species, growing season, fertilizer treatment, irrigation and the stage at which the plants are harvested [4]. Analysing the average annual energy yield (Table 5) for both grass species, it was found that *Dactylis glomerata* with 14,704 JP ha^{-1} had, by 14%, better results than *Lolium perenne* (12,855 JP ha^{-1}). In the case of *Dactylis glomerata,* a significant increase in the annual energy yield compared with the control was reported after the use of compost extract, together with mineral fertilizer, while *Lolium perenne* responded with a higher value to vermicompost applied together with mineral fertilizer.

Table 5. The effect of different treatments on the annual energy yield of *Dactylis glomerata* and *Lolium perenne* (JP ha^{-1}).

	Fertiliser Effect								Means
	0	NPK	CE	VE	HE	CE + NPK	VE + NPK	HE + NPK	
				Means within species					
Dactylis glomerata	9970 ^{Ab}	13,871 ^{Ab}	13,572 ^{Ab}	14,606 ^{Aab}	12,519 ^{Ab}	18,658 ^{Aa}	17,614 ^{Aab}	16,820 ^{Aab}	14,704 ^{A}
Lolium perenne	9432 ^{Ab}	13,912 ^{Aab}	10,751 ^{Ab}	11,643 ^{Ab}	10,775 ^{Ab}	15,583 ^{Aab}	15,951 ^{Aa}	14,791 ^{Aab}	12,855 ^{B}
				Means within growing seasons					
2019	9771 ^{Ab}	15,171 ^{Aab}	13,711 ^{Ab}	14,782 ^{Aab}	12,989 ^{Ab}	18,553 ^{Aa}	18,312 ^{Aa}	16,591 ^{Aab}	14,985 ^{A}
2020	10,199 ^{Ab}	15,879 ^{Aab}	13,082 ^{Ab}	13,775 ^{Ab}	11,481 ^{Ab}	19,850 ^{Aa}	18,639 ^{Aa}	17,160 ^{Aab}	15,008 ^{A}
2021	9133 ^{Ab}	10,624 ^{Bab}	9693 ^{Aab}	10,816 ^{Aab}	10,472 ^{Aab}	12,960 ^{Bab}	13,396 ^{Bab}	13,667 ^{Aa}	11,345 ^{B}
Mean	9700 ^{c}	13,892 ^{b}	12,162 ^{bc}	13,124 ^{b}	11,647 ^{bc}	17,121 ^{a}	16,782 ^{a}	15,806 ^{ab}	
				Means within harvests					
I	3427 ^{Ac}	4641 ^{Ab}	4102 ^{Abc}	4562 ^{Abc}	3982 ^{Abc}	5873 ^{Aa}	5711 ^{Aab}	5327 ^{Aab}	4703 ^{AB}
II	3281 ^{Ac}	4682 ^{Ab}	4301 ^{Abc}	4410 ^{Abc}	4028 ^{Abc}	5941 ^{Aa}	5829 ^{Aab}	5363 ^{Aab}	4729 ^{A}
III	2952 ^{Ab}	4535 ^{Aab}	3851 ^{Ab}	4087 ^{Ab}	3515 ^{Ab}	5269 ^{Aa}	5218 ^{Aab}	5087 ^{Aab}	4314 ^{B}
Mean	3220 ^{c}	4620 ^{b}	4085 ^{bc}	4353 ^{bc}	3841 ^{c}	5695 ^{a}	5586 ^{a}	5259 ^{a}	

0—Control; NPK—mineral fertiliser; CE—compost extract; VE—vermicompost extract; HE—humus extract; Means in lines marked with the same small letters do not differ significantly; Means in columns marked with the same capital letters do not differ significantly.

The average value of the energy yield was the biggest on units where biological extracts were applied together with mineral fertilizer. That increase was from 63% for NPK applied together with humus extract (to 15,806 JP ha^{-1}) to 76.5% for the NPK applied with compost extract (to 17,121 JP ha^{-1}). There was a statistically significant increase by about 35% in the energy yield on plots with vermicompost (13,124 JP ha^{-1}) by about 35%. For the other two biological materials used on their own, an increase was not significant.

Ciepiela et al. [5] found that the energy yield increased with the dose of nitrogen. The authors recorded a 4320 JP ha^{-1} energy yield on the control unit, while on units with nitrogen at a dose of 60 kg ha^{-1} it was 8363 JP ha^{-1}. As an average for grass species and treatments, the largest yield of feed energy was in the first (14,985 JP ha^{-1}) and second (15,008 JP ha^{-1}) years, and the smallest in the third (11,345 JP ha^{-1}). Significantly lower results in the third year might have been caused by plant aging, which decreased both the amount of protein relative to raw fibre and the yield of dry matter.

The largest statistically significant average energy yield was in the second harvest (4729 JP ha^{-1}), and the smallest in the third (4314 JP ha^{-1}). Differences in quantity between harvests were probably caused by varied weather conditions during the growing seasons of the three-year experiment. Each year, dry periods prevailed before the second harvest, which may have contributed to total protein accumulation without increasing crude fibre content. This had a positive impact on the energy yield. In turn, its low value in the third harvest was probably due to a lower yield of plants.

As it is presented in Figure 3a, the annual energy yield of *Lolium perenne* was at a similar level throughout the experiment (from 11,761 to 13,434 JP ha^{-1}), while for *Dactylis glomerata* it declined from 16,600 JP ha^{-1} in the first year to a significantly lower value of 10,929 JP ha^{-1} in the last year. Higher amounts of the annual energy yield of *Dactylis glomerata* in relation to *Lolium perenne* may be due to its characteristics. Generally, *Dactylis glomerata* in comparison with *Lolium perenne* contains more protein and produces higher dry matter yields, which could have affected its energy yield [23,24]. This fact was confirmed by previous studies conducted in Poland under similar physical and chemical conditions of the soil [18,19,25]. The energy yield of *Lolium perenne* throughout the growing season was at a similar level, from 4356 JP ha^{-1} in the first harvest to 4177 JP ha^{-1} in the third (Figure 3b). A more dynamic situation was in the case of *Dactylis glomerata*, for which the yield in the second harvest (5167 JP ha^{-1}) was statistically significantly greater than in the third (4451 JP ha^{-1}).

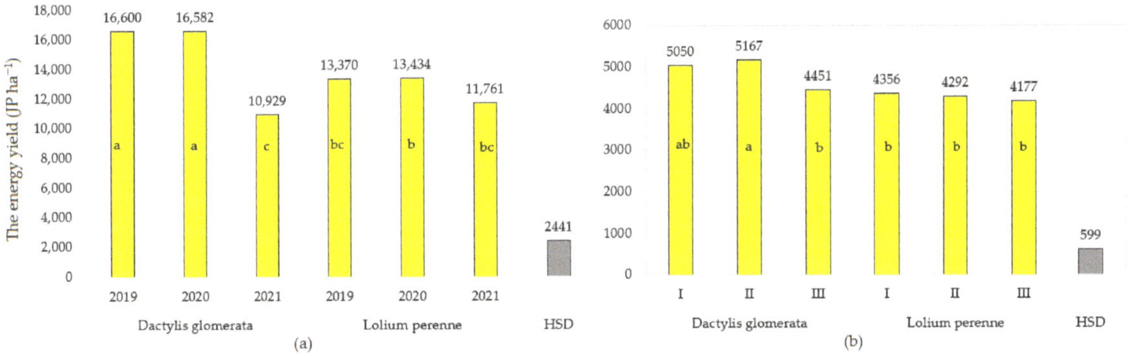

Figure 3. The energy yield of *Dactylis glomerata* and *Lolium perenne* in consecutive (**a**) growing seasons and (**b**) harvests (JP ha^{-1}).

This species had a high dry matter yield in the first and second harvests, but it decreased in the third, leading to a lower energy yield [23]. In a study on the energy yield of *Lolium perenne*, the Research Centre for Cultivar Testing in Słupia Wielka, Poland, observed considerable changes in the values [26]. It ranged from 9062 JP ha^{-1} (the first harvest) to 1858 JP ha^{-1} (the third harvest) in 2015 and from 6133 JP ha^{-1} (the first harvest) to 2355 JP ha^{-1} (the third harvest) in 2016.

4. Conclusions

1. Of the applied biological materials, humic substances applied together with mineral fertilizer had the greatest impact on net energy value and net energy of lactation (NEL).

2. The use of compost extract contributed to a substantial increase in the yield of feed energy. Other biological substances applied together with mineral fertilizer also had a positive impact.
3. *Lolium perenne* feed had a higher net energy of lactation and concentration of net energy than *Dactylis glomerata*; in turn, the latter one had a higher annual yield of feed energy than the former.
4. Weather conditions in various years of research differentiated feed energy values. In 2020, the year with the largest amount of rainfall during most months of the growing period, the feed had the highest value of energy concentration, net energy, and net energy of lactation.
5. Due to the complexity of the environment (soil, climate, plant), it is impossible to provide a universal combination of fertilizers that increases the energy value of forage. Therefore, it is important to carry out practical field experiments that will indicate the optimal fertilizer combinations suitable for the selected region.

Author Contributions: Conceptualization, J.S.; methodology, J.S.; software, M.T.; validation, J.S. and M.T.; formal analysis, M.T.; investigation, J.S.; resources, K.J.; data curation, M.T. and K.J.; writing—original draft preparation, J.S.; writing—review and editing, M.T.; visualization, K.J.; supervision, J.S.; project administration, J.S.; funding acquisition, J.S. All authors have read and agreed to the published version of the manuscript.

Funding: His research was funded by Ministry of Science and Higher Education, Poland, grant number 357/13/S. Publication was co-financed with the project entitled 'Excellent science' program of the Ministry of Education and Science as a part of the contract No. DNK/513265/2021 'Role of agri-culture in implementing concept of sustainable food system "from field to table.

Institutional Review Board Statement: Not applicable.

Informed Consent Statement: Not applicable.

Conflicts of Interest: The authors declare no conflict of interest.

References

1. Kujawiak, R.; Zarudzki, R. DLG system in feeding dairy cows. *Nowocz. Zyw. Zwierząt* **2001**, *6*, 18–19. (In Polish)
2. Jonker, J.S.; Kohn, R.A.; Erdman, R.A. Milk urea nitrogen target concentratsis for lactating dairy cows fed according to national research council recommendations. *J. Dairy Sci.* **1999**, *82*, 1261–1273. [CrossRef]
3. Juszczak, J.; Ziemiński, R. The urea content in milk as an indicator of the protein-energy ratio in the feed dose for dairy cows. *Adv. Agric. Sci.* **1997**, *3*, 73–82. (In Polish)
4. Abas, I.; Ozpinar, H.; Can Kutay, H.; Kahraman, R. Determination of the Metabolizable Energy (ME) and Net Energy Lactation (NEL) Content of Some Feeds in the Marmara Region by In vitro Gas Technique. *Turk. J. Vet. Anim. Sci.* **2005**, *29*, 751–757.
5. Ciepiela, G.A.; Jankowska, J.; Kolczarek, R. Yield of feed units obtained from cocksfoot cultivated in pure sowing and in mixtures with leguminous plants. *Zesz. Nauk. WSA Łomża* **2008**, *37*, 83–88. (In Polish)
6. Vulchinkov, Z.; Katova, A. Forage productivity evaluation in cocksfoot (*Dactylis glomerata* L.) accessions. *Bulg. J. Crop Sci.* **2020**, *57*, 101–108.
7. Kosolapova, T.V.; Tulinov, A.G. Assessment of adaptability parameters of cocksfoot in the conditions of the Komi Republic. *Russ. Agric. Sci.* **2022**, *47*, 562–567. [CrossRef]
8. Zhouri, L.; Kallida, R.; Shaimi, N.; Barre, P.; Volaire, F.; Gaboun, F.; Fakiri, M. Evaluation of cocksfoot (*Dactylis glomerata* L.) population for drought survival and behavior. *Saudi J. Biol. Sci.* **2019**, *26*, 49–56. [CrossRef]
9. Nguyen, T.M.; Meixue, Z.; David, P.; Rowan, W.S. Aerenchyma formation in adventitious roots of tall fescue and cocksfoot under waterlogged conditions. *Agronomy* **2021**, *11*, 2487.
10. Siaudinis, G.; Jasinskas, A.; Karcauskiene, D.; Sarauskis, E.; Lekaviciene, K.; Repsiene, R. The dependence of cocksfoot productivity of liming and nitrogen application and the assessment of qualitative parameters and environmental impact using biomass for biofuels. *Sustainability* **2020**, *12*, 8208. [CrossRef]
11. Reed, K.F.M. Improving the adaptation of perennial ryegrass, tall fescue, Phalaris, and cocksfoot for Australia. *N. Z. J. Agric. Res.* **1996**, *39*, 457–464. [CrossRef]
12. Butler, T.J.; Muir, J.P.; Provin, T.L. Phosphorus fertilization of annual ryegrass and comparison of soil phosphorus extractants. *J. Plant Nutr.* **2007**, *30*, 9–20. [CrossRef]

13. Kotecki, A. Climate change for the European Green Deal. In Proceedings of the Scientific Conference: The Role of Agricultural Sciences in the Implementation of the Concept of a Sustainable Food System "from Farm to Fork", Chełm, Poland, 7–8 June 2022. (In Polish).
14. Sosnowski, J. The value of production, energy and food of *Festulolium brauni* (K. Richt.) A. Camus microbiologically and mineral suppled. *Fragm. Agron.* **2012**, *29*, 115–122. (In Polish)
15. Sosnowski, J.; Jankowski, K. Effect of soil medium amendment on chemical composition and digestibility of *Lolium multiflorum* Lam. *Sci. Nat. Technol.* **2013**, *7*, 15. (In Polish)
16. Truba, M.; Wiśniewska-Kadżajan, B.; Jankowski, K. The influence of biology preparations and mineral fertilization NPK on fiber fractions content in *Dactylis glomerata* and *Lolium perenne. Fragm. Agron.* **2017**, *34*, 107–116. (In Polish)
17. Truba, M.; Wiśniewska-Kadżajan, B.; Sosnowski, J.; Malinowska, E.; Jankowski, K.; Makarewicz, A. The effect of soil conditioners on cellulose, hemicellulose, and the ADL fibre fraction concentration in *Dactylis glomerata* and *Lolium perenne. J. Ecol. Eng.* **2017**, *18*, 107–112. [CrossRef]
18. Truba, M.; Jankowski, K.; Wiśniewska-Kadżajan, B.; Malinowska, E. The effect of soil conditioners on the content of soluble carbohydrates, digestible protein and the carbohydrate-protein ratio in *Lolium perenne* and *Dactylis glomerata. Environ. Prot. Nat. Resour.* **2018**, *29*, 1–8. [CrossRef]
19. Truba, M.; Jankowski, K.; Wiśniewska-Kadżajan, B.; Sosnowski, J.; Malinowska, E.; Barszczewski, J. The effects of soil conditioners on total protein and crude fiber concentration in selected grass species. *Appl. Ecol. Environ. Res.* **2018**, *16*, 2729–2739. [CrossRef]
20. Schad, P.; van Huyssteen, C.; Micheli, E. International soil classification system for naming soils and creating legends for soil maps: World Soil Resources Reports. In *World Reference Base for Soil Resources 2014*; World Soil Resources Reports No. 106; FAO: Rome, Italy, 2014.
21. Wiśniewska-Kadżajan, B. Effect of mushroom substrate on the feed quality from the permanent meadow. *J. Ecol. Eng.* **2014**, *15*, 45–49.
22. Kolczarek, R.; Jankowska, J.; Ciepiela, G.A.; Jodełka, J. The nutritional value of sward from a permanent meadow depending on the type of fertilization. *Wiad. Mel. i Łąk.* **2008**, *4*, 181–183. (In Polish)
23. Downing, T.; Gamroth, M. *Nonstructural Carbohydrates in Cool-Season Grasses*; Special Report 1079; Oregon State University: Corvallis, OR, USA, 2007.
24. Tilvikiene, V.; Kadziuliene, Z.; Dabkevicius, Z.; Sarunaite, L.; Slepetys, J.; Pociene, L.; Slepetiene, A.; Ceceviciene, J. The yield and variation of chemical composition of cocksfoot biomass after five years of digestate application. *Grassl. Sci. Eur.* **2014**, *19*, 468–470.
25. Jankowski, K.; Truba, M.; Wiśniewska-Kadżajan, B.; Sosnowski, J.; Malinowska, E. The effect of soil conditioners on yield of dry matter, protein, and sugar in grass. *Acta Agroph.* **2018**, *25*, 45–58. [CrossRef]
26. Centralny Ośrodek Badania Odmian Roślin Uprawnych. *Synthesis of the Results of Registry Experiments, Poaceae (Grass) 2013–2016*; Centralny Ośrodek Badania Odmian Roślin Uprawnych: Słupia Wielka, Poland, 2017; 142p. (In Polish)

agriculture

MDPI

Article

Diversification of Agricultural Output Intensity across the European Union in Light of the Assumptions of Sustainable Development

Aneta Zakrzewska and Anna Nowak *

Department of Economics and Agribusiness, University of Life Sciences in Lublin, Akademicka 13, 20-950 Lublin, Poland
* Correspondence: anna.nowak@up.lublin.pl

Abstract: The strength of the bond between agriculture and the natural environment is measured by output intensity. This work aimed to evaluate the diversity of agriculture across the European Union in terms of agricultural output intensity from the perspective of the assumptions of the concept of sustainable development. Surveys were conducted using selected indicators based on data derived from EUROSTAT, FAOSTAT, and FADN from 2010–2019. The adopted indicators were used for developing a ranking of member states according to output intensity, which, in confrontation with the level of agricultural efficiency, can form a basis for an individual approach to the development strategies of respective member states. Their findings imply that, in the analyzed period, differences in output intensity among member states declined insignificantly. From 2010 to 2019, most countries forming the so-called 'old 15' featured higher output intensity than new member states. The Netherlands and Malta recorded the highest cost of intermediate consumption per 1 ha of utilized agricultural area. By contrast, agricultural production was the least intensive in Bulgaria. Land productivity was also very strongly variable. The difference between the old and new member states was clearly marked. Dutch agriculture reached the highest land productivity from 2010 to 2019, where agricultural production levels per 1 ha were five times higher than on average in the European Union.

Keywords: agriculture; European Union; output intensity; sustainable development

1. Introduction

Agriculture is a sector of the economy that has special links with the natural environment. On the one hand, its efficiency depends on the environmental resources; on the other hand, agricultural production often takes advantage of the natural environment (e.g., degrades soil and water quality and reduces biological diversity), which is detrimental to environmental sustainability [1,2]. Agriculture is a source of half of the local emissions of greenhouse gases other than CO_2 in the European Union (EU), and one-third of water intake is utilized for its needs [3]. In addition, Zegar [4] noted that agriculture significantly impacts the climate, accounting for nearly one-third of anthropogenic changes and, to some extent, for the loss of biological diversity. The industrial agriculture model, which became very efficient but generated global negative environmental and social effects, proved to be a particular burden to the natural environment [5]. Since the paradigm of European agriculture after the period of its industrialization needed to be altered, the sustainability of agriculture was deemed a priority direction of development reflected by the Common Agricultural Policy of the European Union [6,7]. A review of the literature led to a conclusion that, despite the great popularity of the notion of sustainable development, it has not been precisely defined [8,9]. The model of sustainable agriculture can be identified with a harmonious link among the efficient production of goods and services (economic function), the management of natural resources (environmental function), and improved living standards in rural areas (social function) [7,10,11]. The economic, environmental, and social dimensions of sustainable agriculture are, to some extent, complementary. Prosperous and efficient agriculture is capable of investing in environment-friendly production, and environment-friendly production and low prices of agricultural products are beneficial from the point of view of the whole society. However, these three dimensions of sustainable agriculture can be conflicting as intensive agricultural production degrades the natural environment. Therefore, the objectives of CAP have been reoriented towards a model of agriculture and regulating mechanisms having a beneficial impact on the natural environment [12]. The intensification of production itself is not a negative phenomenon since it supports achieving economic and social goals. However, excessively intensive production, through its adverse impact on the natural environment, imposes limitations on achieving the environmental goals of sustainable development. Sustainable agriculture is a global, dynamic process within these three areas, occurring at five levels: field, agricultural holding, local community, national, and international levels [13,14]. Czyżewski and Stanisławski [15] underlined that the development of agriculture consistent with the paradigm of sustainable development became particularly important in industrialized countries where the development of the agricultural sector previously followed the industrial model. However, it proved unreliable in the long run. Surveys dedicated to sustainable agriculture suggest that sustainability reduces certain social costs generated by industrial agriculture. Furthermore, the higher the socio-economic stability of respective countries is, the more often they demonstrate green attitudes and participate in agri-environmental programs [16,17].

Since agriculture is deemed the main keeper of the natural environment, it should become a leader in the change towards sustainable development through promoting innovative technologies and governance models [18,19]. The care for the quality of nature and natural resources is not only a requirement of civilization but also a prerequisite for agricultural production [20]. The impact of agriculture on the natural environment depends on its output intensity. In the age of industrialization, in order to meet a high requirement for food, measures to the extent of agriculture aimed to intensify agricultural production. This was accomplished by increasing the capital expenditure per land resource unit in order to achieve an increase in production [21]. Keys and McConnell [22] defined the intensification of agriculture as a process of increasing input per area unit or increasing output per area unit. Agricultural output intensity can be described by various indicators referring to structural and organizational, natural and agrotechnological, and economic and organizational conditions [23,24]. The most popular indicators of agricultural output intensity are labor and capital inputs per unit of utilized agricultural area (UAA) [25]. Many scientific papers also refer to a measure of land productivity, that is, crop yield [25–27]. Ruiz-Martinez et al. [28] reviewed studies and indicators that refer to assessing agricultural output intensity, grouping them as input-oriented and output-oriented measures. The latter includes land productivity, an indicator of the relationship between the value of production and the utilized agricultural area. Regional surveys also measure agricultural output intensity regarding an area needed to produce one production unit [29]. By contrast, Levers et al. [27] attempted to design agricultural intensity patterns across Europe. They analyzed the spatial conditions for change in the intensity of agricultural production using a set of biophysical and socio-economic variables. They demonstrated that higher crop yield was usually associated with an increased use of fertilizers, high soil quality, and high labor efficiency. At present, intensification factors pose no threat to the natural environment; the quality of products and consumer health are particularly important [30]. According to Zegar [4], agricultural intensification compliant with the need for sustainability, that is, ecological intensification, should be promoted and supported. This is reflected by the directions of development set out in the Common Agricultural Policy. Thus, agricultural intensification is not only a measure aimed at increasing food production but also a huge challenge to global ecosystems [31].

The challenges to the development of present-day agriculture include living up to the competitors. In addition, a fundamental development dilemma arises: how to reconcile sustainable development with intensifying competition, that is, how to achieve a competitive advantage [21]. The quality of agricultural production space and the level of output intensity impact both the level of farmers' income and the quality of agricultural products. Therefore, taking care of the fertility and prolificacy of agricultural land and moderating the intensity of production processes are important challenges. It is obvious that, in view of the growing global population, it is necessary to increase agricultural production [32–34]. Therefore, it is important to evaluate the diversity of output intensity, on the one hand, showing the degree of impact of agriculture on the natural environment [35] and, on the other hand, possible options of increasing the production output. The role of such surveys is also due to the fact that, in the coming decades, the competition between food production and other uses of water and land will probably increase [36]. The need for evaluating the intensity of agricultural production in the European Union is dictated by the fact that agriculture varies among respective members in terms of production potential, efficiency of its use, and area structure. Research conducted by many authors implies that new member states have a lower potential of agricultural development, including an agrarian structure not supporting high production efficiency [37,38]. This is due to, among other things, the central planning of the economy and limited access to production factors in most of them after World War II. An exception is the agriculture in Slovakia and Czechia as these countries have a more favorable agrarian structure. In the countries of Western Europe, land was subject to concentration. This process, in contrast to collectivization, was not controlled by political decisions but was forced by the market situation. Such an agrarian structure determines manufacturing technologies and related output intensity. Large and very large agricultural holdings tend to apply development strategies based on specialization and intensification, while small ones are more often susceptible to diversification [39,40]. Thus, working on development strategies for sustainable agriculture, structural characteristics should be taken into account next to differences in output intensity among EU member states. As regards the differentiation of agriculture in the European Union and the absence of a uniform approach to output intensity, a research gap in the comprehensive assessment of agricultural output intensity can be identified in all EU member states, in particular in the long term. Available studies most often refer to single, isolated measures or to selected EU member states only. Incorporating a wide range of diagnostic features describing agricultural output intensity in all member states in the assessment constitutes this paper's added value. In addition, such assessments are rarely made in light of sustainable development assumptions.

In view of the relationship of agriculture with the natural environment, this study aimed to evaluate the diversity of agriculture in the member states of the European Union in terms of agricultural production diversity. This assessment was performed in the context of sustainable development assumptions. This work is structured as follows. The following section contains a description of the research methods. In Section 3, we present a ranking of EU member states according to output intensity, confronted with the level of land productivity. An important element of research is the thorough analysis of output intensity, taking into account additional features expressing the relationship among production factors and, in particular, labor and land. The last section presents conclusions from the analyses.

2. Materials and Methods

The surveys were based on selected indicators designed using data from the EUROSTAT, FAOSTAT, and FADN database. The analysis covered the years 2010–2019, which allowed us to determine changes in the intensity of production over a decade. Twenty-eight member states of the European Union were the subjects of the study. The United Kingdom is no longer a member of the EU, but, in the examined period, it remained within its structures.

In general, intensity refers to the degree of any phenomenon or human activity involved in the process of production. It reflects the actual expenditure on the process of production. Agricultural output intensity is a measure of the degree of utilization of land by means of other factors. Therefore, the basic measure of output intensity was the cost of agricultural production per 1 ha of utilized agricultural area. To this end, the costs of intermediate consumption were used that, according to the methodology of FADN and EUROSTAT, include direct costs (including products made and used in the process of production on a farm) and general economic costs accompanying the operations in the accounting year [41]. Thus, output intensity and the impact of agriculture on the natural environment increases with the increase in the level of intermediate consumption per 1 UAA. In addition, output intensity was evaluated based on the consumption of nitrogen, potash, and phosphorus per 1 ha of utilized agricultural area and the cost of plant protection products, herbicides, insecticides, and pesticides per 1 ha UAA. The characteristics were selected based on their substantive merits and a review of the reference literature. At the second stage of research, based on the measured values of output intensity indica-tors, a synthetic index of output intensity for respective member states of the European Union in 2010–2019 was designed. It was assumed that the total output intensity increased at higher level of respective indicators. To this end, each indicator was regarded as a stimulant and normalized as follows:

$$z_{ij} = \frac{x_{ij} - \min(x_{ij})_i}{\max(x_{ij})_i - \min(x_{ij})_i} \rightarrow (i = 1, 2, \ldots, n;\ j = 1, 2, \ldots, m) \quad (1)$$

where mini$\{x_{ij}\}$ is the minimum value of feature j, maxi$\{x_{ij}\}$ is the maximum value of feature j, and i is the object (country).

As a result of this operation, values of respective indicators fell within the range [0, 1]. Values closer to 1 mean that the specific variable (indicator) is better and, to the contrary, values closer to 0 mean that the specific indicator is worse. Next, synthetic indices of output intensity were calculated for respective EU member states. The indicator was an arithmetic mean consisting of standardized partial indicators [42]. The synthetic measure was used for developing a ranking of member states according to output intensity, which, in confrontation with the level of agricultural efficiency, can form a basis for an individual approach to the sustainable development strategies of respective member states of the EU.

The analysis also covered the structure of utilized agricultural area managed by agricultural holdings featuring low, medium, and high input intensities per 1 ha. According to FADN, farms are classified into three intensity categories according to the estimated input per 1 hectare of utilized agricultural area. Inputs taken into account include fertilizers, pesticides, other crop protection chemicals, and purchased feed. A low-intensity farm is one for which inputs are lower or equal to the value of intensity associated with the 33rd quantile (Q33). A high-intensity entity is one with inputs exceeding the value of intensity determined by the 66th quantile (Q66). Farms with the level of inputs exceeding the value of intensity described by Q33 but lower than the value described by Q66 are classified as medium-intensity ones [3].

The evaluation also covered labor input per 100 hectares (ha) of UAA and land productivity measured as the production value per 1 ha of utilized agricultural area. In addition, the relationship between the level of intermediate consumption per 1 ha of UAA and land productivity was examined using Pearson's correlation coefficient. A comparison of the level of output intensity with land productivity will be the basis for identifying countries in which, due to the low level of expenditure, further intensification processes are not contrary to the concept of sustainable development.

In addition, the significance of differences in indicators describes output intensity between new (EU-15) and old (EU-13) member states of the EU. The analysis was preceded by the verification of the assumptions of the Student's t-test for the normality of distribution (Shapiro–Wilk test) and for the homogeneity of variance (Levene's test). In the case of variables for which at least one assumption was not met, the Mann–Whitney test was applied during analysis. The analysis used Excel and a statistical package, Jamovi.

3. Results and Discussion

The literature shows a big diversity across the European Union both in terms of employment, costs of labor in agriculture, productivity, and production [43,44]. In this study, the output intensity level was expressed as the cost of production per 1 ha of utilized agricultural area. Figure 1 presents their average values from 2010 to 2019 in the member states of the EU in an ascending order.

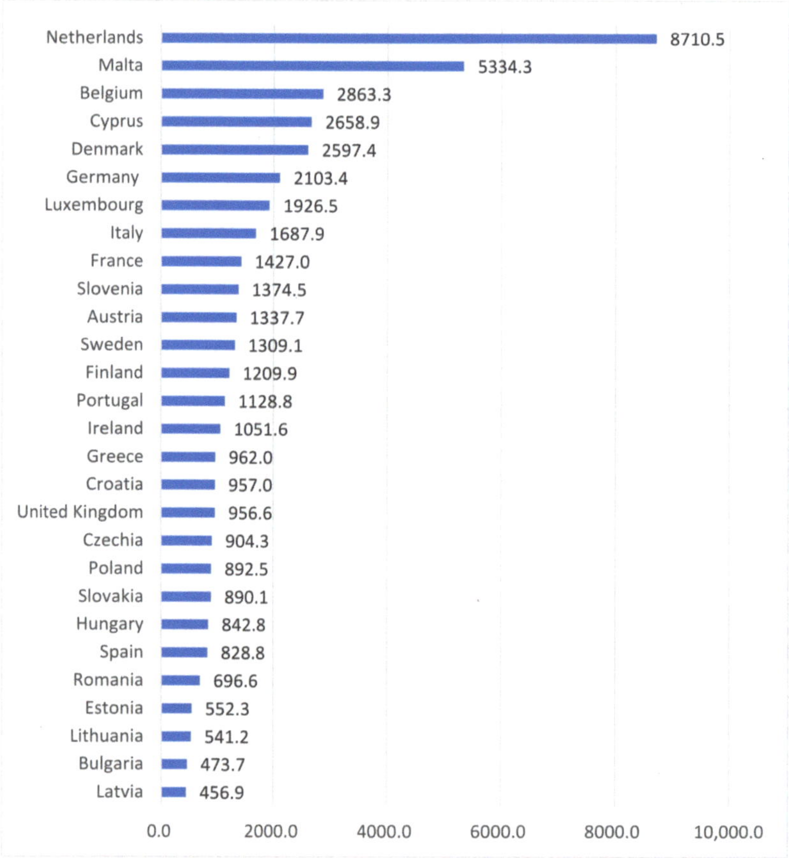

Figure 1. Intermediate consumption costs per hectare of utilized agricultural area (UAA) in the European Union countries, 2010–2019 (EUR per 1 ha). Source: own calculation The average value of the analyzed indicator in 28 member states of the EU from 2010 to 2019 was EUR 1667. Its level was higher than the EU mean in eight countries, and 13 did not exceed EUR 1000. The old and new member states can observe a clear difference in output intensity. In the analyzed period, the average value of intermediate consumption per 1 ha of utilized agricultural area in 13 new member states of the EU amounted to EUR 1275, while the mean for the EU-15 was EUR 2006.70. In countries such as Bulgaria, Latvia, Estonia, and Lithuania, production in the analyzed years was the least intensive. Poland incurred relatively low costs per 1 ha of UAA, accounting for 53.5% of those incurred in the EU (Figure 1). In each of the examined years, the coefficient of variation in the analyzed feature exceeded 100%, which means it was strongly diversified. At the same time, it can be seen that in 2010 the diversity was higher (105.4%) than in 2019 (102.6%).

It should also be highlighted that, from 2010 to 2019, the level of intermediate consumption per 1 ha of UAA changed. In Poland, the analyzed indicator increased by 24.7%. The cost per 1 ha of UAA also increased by more than 20% in Estonia, Romania, Latvia, and Austria. In contrast, the highest extensification was observed in Belgium and Cyprus (Figure 2).

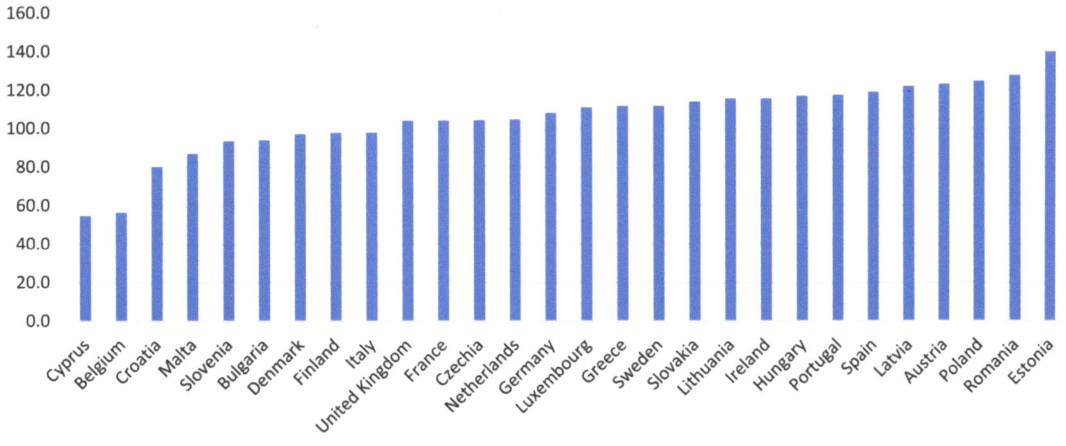

Figure 2. The dynamics of change in intermediate consumption per 1 ha of UAA from 2010 to 2019 (2010 = 100).

Fertilization is a fundamental yield-forming factor and the consumption of fertilizers is one of the indicators for assessing the intensity of management [45,46]. The level and dynamics of using fertilizers are definitely determined by the economic development status of the respective country [47]. Therefore, the analysis covered the consumption of nitrogen, potash, and phosphorus per 1 ha of utilized agricultural area and the cost of plant protection products, herbicides, insecticides, and pesticides per 1 ha UAA (Table 1). The presented data imply that, in the study period, Luxembourg, the Netherlands, and Belgium were among the leaders in the consumption of nitrogen for agricultural production, consuming more than 200 kg N/ha. As regards phosphorus, Croatia, Ireland, and Cyprus stand out. By contrast, Ireland and Belgium were the biggest consumers of potash. It should be noted that, next to the economic effects, the intensive growing of crops using a high dosage of nitrogen, phosphorus, and potash leads to risks for the natural environment. The concept of sustainable development searches for a compromise between economic and environmental criteria. It assumes that permanent and fair development relies on, among other things, a combination of the laws of ecology and economics in the decision-making process. The authors of various reference literature discuss issues related to the intensity of fertilization and the chemical protection of plants in the context of environmental protection and a nature conservation policy [48,49]. The excessive use of nitrogen and phosphorus fertilizers is particularly hazardous at low levels of lime fertilizers [50]. Meanwhile, studies by Zalewski [51] implied that, from 2010 to 2018, the member states of the European Union increased the total value of fertilizers and plant protection products used. The value of the analyzed means of production for agriculture per 1 ha UAA, and their share in intermediate consumption, also increased. Analyzing the cost of plant protection products, herbicides, insecticides, and pesticides per 1 ha, it can be noted that it is highly differentiated across the EU. The Netherlands, Belgium, and Cyprus stand out clearly, with the abovementioned cost amounting to EUR 211.3, EUR 165.8, and EUR 154 per 1 ha UAA, respectively. By contrast, countries featuring a relatively low cost of plant protection products were Ireland, Romania, and Latvia (Table 1).

Table 1. Consumption of mineral fertilizers and plant protection products, herbicides, insecticides, and pesticides per 1 ha in the EU member states from 2010 to 2019.

Member State	Nitrogen (N) Consumption in Mineral Fertilizers	Phosphate (P_2O_5) Consumption in Mineral Fertilizers	Potash (K_2O) Consumption in Mineral Fertilizers	Cost of Using Plant Protection Products, Herbicides, Insecticides, and Pesticides
	kg/ha	kg/ha	kg/ha	EUR/ha
Austria	78.7	20.3	23.7	45.1
Belgium	201.2	25.3	75.5	165.8
Bulgaria	90.1	16.8	8.4	31.3
Croatia	103.9	51.2	41.4	60.5
Cyprus	66.1	40.5	27.7	154.0
Czechia	131.1	18.4	11.6	59.9
Denmark	86.1	13.9	27.2	67.1
Estonia	52.6	11.8	16.7	23.2
Finland	63.5	11.4	15.5	36.0
France	111.1	20.9	23.2	105.2
Germany	134.4	21.5	35.4	93.6
Greece	53.5	16.5	13.3	40.0
Hungary	78.9	18.5	18.3	68.6
Ireland	163.0	47.9	107.4	15.2
Italy	62.5	18.6	13.7	60.8
Latvia	57.5	18.4	20.9	28.3
Lithuania	71.5	20.3	26.8	39.6
Luxembourg	211.8	17.5	15.9	60.9
Malta	55.1	9.1	12.7	53.4
Netherlands	206.2	14.0	41.0	211.3
Poland	96.1	31.1	44.3	65.8
Portugal	59.5	21.7	18.4	31.5
Romania	38.7	15.0	5.4	19.2
Slovakia	84.1	15.9	11.7	61.1
Slovenia	115.9	38.4	46.8	43.1
Spain	57.8	23.9	21.3	38.8
Sweden	68.7	10.8	12.3	31.9
United Kingdom	167.4	31.3	43.8	58.1

Source: own calculation.

At the next stage of research, a synthetic measure of output intensity was calculated for respective member states of the EU based on standardized indicators given in Table 1. Figure 3 presents the ranking of member states, implying that agricultural output intensity is the highest in the Netherlands, Belgium and Ireland. The ranking is closed by Romania featuring the weakest impact of agriculture on the natural environment.

In the contemporary economic theory, the prevailing view is the necessity of orienting innovation towards input-saving production techniques [52]. Therefore, innovativeness should aim to reduce the use of production resources in the production process. In particular, this refers to agriculture being the main keeper of the natural environment. Thus, it is desirable to disseminate technologies that simultaneously reduce the costs per unit of agricultural production and the negative environmental impact of production. In the European Union, an important instrument in this respect is the Common Agricultural Policy (CAP), with objectives that have evolved to support the sustainable management of the natural resources, reduce adverse impacts of agricultural production on the natural resources, and prevent climate change [53]. According to the European Commission, a reduction in the level of pesticides used in agriculture is essential to the natural environment, consumer health, and the economy. Currently, the production of many fertilizers and pesticides relies on limited fossil resources [54]. Therefore, in the fertilizers' sector, transformation into a circular economy (CE) focusing on the efficient utilization of resources is justified. References to the fertilizers' sector in the EU and the consumption of fertilizers in agriculture are also present in the new EU growth strategy, that is, in the European Green Deal. It provides for a projected considerable reduction in the volume of wastes and reducing the impact of agricultural fertilizers on the natural environment.

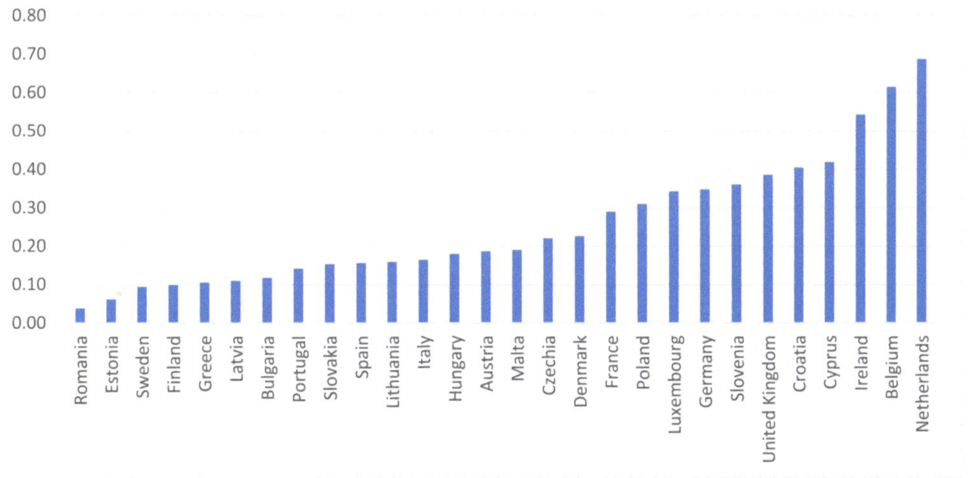

Figure 3. Ranking of EU member states according to agricultural output intensity from 2010 to 2019.

In 2019, the utilized agricultural area in the European Union managed by low input intensity farms accounted for 31.6% of the overall utilized agricultural area, whereas the area at the disposal of farms with a medium and high level of input corresponded to 27.4% and 41%, respectively (Table 2). Countries with the largest share of the utilized agricultural area remaining at the disposal of farms featuring the highest output intensity on the scale of the specific country in 2019 were Romania, Estonia, Slovakia, Bulgaria, and Czechia. At the same time, it can be noted that in each of these countries, the percentage of UAA managed by agricultural holdings with high input intensity increased compared to that noted in 2010. This means that output intensified in those countries. However, considering the absolute value of inputs, intensification in countries with low output intensity does not mean the same for the environment as intensification in a country where this level is high. Belgium and Denmark, where the production has been clearly extensive but with a considerably higher level of inputs per 1 ha than on average in the EU, are worth noting. Next to output intensity, the intensity of production organization, expressed, for instance, as the stocking density per ha UAA, is also worth noting. This indicator reflects the environmental pressure of animal husbandry. According to EUROSTAT [55], in 2016, the stocking density of farm livestock in the EU-28 was 0.8 livestock units (LSU) per one hectare of utilized agricultural area (UAA). This level was slightly higher than in the previous EU survey of agricultural holdings conducted in 2013. A member state with the highest farm livestock density, amounting to 3.8 LSU/ha, was the Netherlands, followed by Malta and Belgium. These three countries noted the highest grazing livestock density. The lowest total farm livestock density among the member states was observed in Bulgaria (0.2 LSU/ha). However, from 2013 to 2016, this indicator noted the highest increase among the member states (by 11.1%). Kopiński [56] underlined that the organization of production in animal husbandry becomes more extensive (specialized) and simultaneously more intensive (concentrated), which can increase environmental pressure on areas where animal husbandry is very concentrated, leading to, among other things, the deterioration of the quality of surface and ground water.

Table 2. The structure of utilized agricultural area managed by agricultural holdings according to the intensity of production inputs in the member states of the EU in 2010 and 2019 (%).

Country	High Input Intensity		Medium Input Intensity		Low Input Intensity	
	2010	2019	2010	2019	2010	2019
Belgium	31.4	15.1	34.9	22.0	33.7	62.9
Bulgaria	45.1	72.0	25.2	14.8	29.7	13.2
Czechia	40.2	45.7	27.6	26.6	32.2	27.7
Denmark	34.1	27.3	36.2	31.9	29.7	40.8
Germany	37.7	42.5	34.1	29.1	28.2	28.4
Estonia	52.1	56.0	21.5	13.7	26.4	30.3
Ireland	32.6	40.0	31.4	28.8	36.0	31.2
Greece	32.1	28.8	32.5	31.5	35.4	39.7
Spain	36.3	40.8	29.0	30.4	34.7	28.8
France	33.7	37.4	33.9	29.7	32.4	32.9
Croatia	33.5 *	52.9	34.1 *	23.8	32.4 *	23.3
Italy	33.7	33.8	30.8	31.0	35.5	35.2
Cyprus	34.2	17.0	33.6	31.3	32.2	51.7
Latvia	35.7	52.5	31.8	19.5	32.5	28.0
Lithuania	31.3	46.1	32.6	21.0	36.1	32.9
Luxembourg	39.7	50.4	34.5	21.1	25.8	28.5
Hungary	39.9	40.2	33.9	33.9	26.2	25.9
Malta	31.4	28.2	33.6	30.3	35	41.5
Netherlands	34.4	49.1	32.7	28.3	32.9	22.6
Austria	34.2	51.0	35.5	25.7	30.3	23.3
Poland	31.4	30.6	34.1	34.7	34.5	34.7
Portugal	32.7	40.5	30.0	30.8	37.3	28.7
Romania	33.0	72.3	34.0	20.1	33.0	7.6
Slovenia	34.4	31.8	35.9	32.4	29.7	35.8
Slovakia	48.7	54.9	24.3	20.5	27.0	24.6
Finland	30.3	26.4	30.8	36.8	38.9	36.8
Sweden	34.2	33.0	31.8	34.8	34.0	32.2
United Kingdom	32.2	32.3	33.8	33.1	34.0	34.6
EU-28	35.7	41.0	31.9	27.4	32.4	31.6

* Data for 2013.

Recently, the number of workers employed in agriculture has decreased and that in the services sector has increased in Poland and the majority of developed economies. A significant element of the analysis of the economic situation of agriculture is the relationship among production factors. An indicator describing the type of production techniques used is the number of workers per 100 ha of utilized agricultural area. From 2010 to 2019, that ratio ranged from 1.7 AWU in the United Kingdom to 43.1 AWU in Malta. Poland had, on average, 13 workers per 100 ha (Figure 4). The surveys support the view that, in Poland, excessive employment in agriculture and a negative labor–land ratio still persist. Parzonko [57] underlined that this results from an adverse agrarian structure, the so-called covert unemployment in agriculture. In the examined period, the analyzed ratio declined in most member states (except the Netherlands, Austria, Malta, Slovenia, the United Kingdom, and Greece).

The significance of differences in agricultural output intensity indicators between old (EU-15) and new (EU-13) member states was analyzed using the non-parametric Mann-Whitney test comparing the significance of differences between the two groups. The level of significance was adopted as $\alpha = 0.05$ (Table 3).

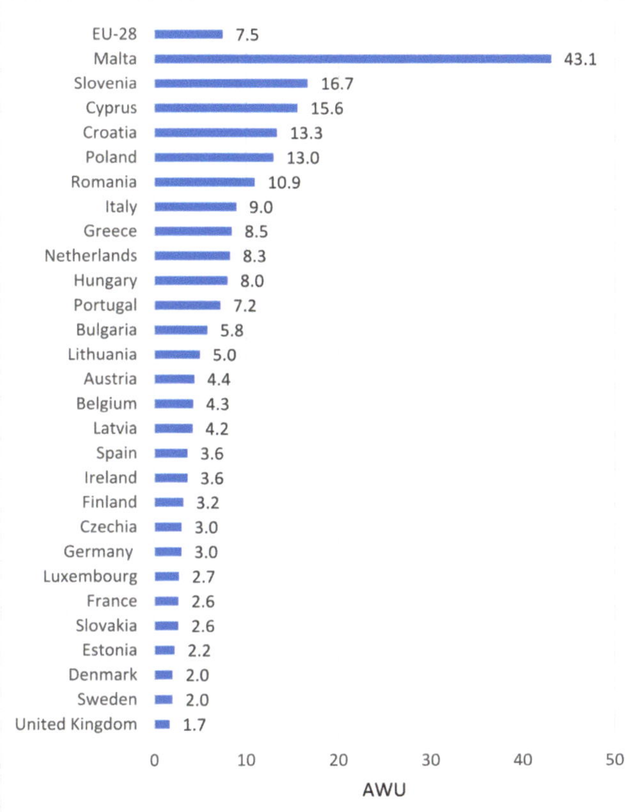

Figure 4. The number of full-time equivalents in agriculture per 100 ha of UAA in the member states of the EU from 2010 to 2019 (AWU per 100 ha of UAA).

Table 3. Assessment of the significance of differences between EU-15 and EU-13 groups (Mann-Whitney test results).

Variables					95% CI	
	U		p	rrb	Lower	Upper
Intermediate consumption costs per hectare of UAA	41.00	**	0.008	−0.58	−0.80	−0.22
Nitrogen (N) consumption in mineral fertilizers per ha	69.00		0.201	−0.29	−0.63	0.13
Phosphate (P2O5) consumption in mineral fertilizers per ha	96.00		0.964	−0.02	−0.42	0.40
Potash (K2O) consumption in mineral fertilizers per ha	74.00		0.294	−0.24	−0.59	0.19
The number of full-time equivalents in agriculture per 100 ha of UAA	145.00	*	0.029	0.49	0.10	0.75
Cost of using plant protection products, herbicides, insecticides, and pesticides per ha	82.00		0.496	−0.16	−0.53	0.27

* $p < 0.05$, ** $p < 0.01$, *** $p < 0.001$.

The results of the Mann–Whitney test point to significant differences between the examined groups of countries as regards the following two variables:

- Intermediate consumption costs per hectare of UAA: the observed effect size (rrb) was very big and the results in the EU-15 group were significantly higher than in the EU-13 group.
- The number of full-time equivalents in agriculture per 100 ha of UAA: the observed effect size (rrb) was very big and the results were significantly higher in the EU-13 category than in the EU-15 category.

Productivity is a measure of efficiency in agriculture, and land productivity is an expression of the value of agricultural output per 1 ha of utilized agricultural area. Analyzing the productivity of production factors in agriculture in the EU, Tarnowska [58] found that, in 'new' member states of the EU, the involvement of land inputs in achieving agricultural output from 2005 to 2012 was less productive (in terms of value) than in the EU-15. Data in Table 3 imply that land productivity in the EU was diverse and increased in most member states from 2010 to 2019. According to Tarnowska [58], this increase was determined by biological, chemical, technological, and organizational developments. Simultaneously, clear disparities in land productivity can be observed between old and new member states. This reflects differences in output intensity, as Pearson's correlation coefficient for these two variables amounted to 0.984. The average value of agricultural production per 1 ha of utilized agricultural area in the EU from 2010 to 2019 was EUR 2779.70 (Table 4). Out of the countries that joined the EU in or after 2004, Cyprus was the only one with a productivity index exceeding its mean value in the EU. In the examined years, in Poland, the value of agricultural output per 1 ha of UAA was on average EUR 1554.90 and the dynamics were 115.8. Zhang et al. [59] proved that the misallocation of resources in agriculture was the main reason for decreased productivity in countries/regions featuring a lower level of development. Thus, in countries with a lower output intensity, there is a potential for growth in land productivity, which should, however, be accompanied by sustainable production intensification taking the environmental objectives of sustainable development into account.

Figures 5 and 6 show the position of respective countries separately for the old and new member states, depending on the cost of intermediate consumption per 1 ha of UAA and land productivity. By contrast, the size of the balls refers to the percentage of the EU's agricultural production. To improve the readability of charts, the outliers were removed. For the EU-15, it was the Netherlands, and for the EU-13, Malta and Cyprus. The position of the Netherlands standing out from that of other countries is due to a very high output intensity and land productivity. In addition, from 2010 to 2019, it accounted for 6.7% of the EU's agricultural output value. Belgium and Denmark are also far from the cluster of other countries, although their shares in the EU's output were considerably lower and amounted to 2.2% and 2.8%, respectively. Additionally, Germany, Italy, and France are three countries that are worth noting for their high significance to agricultural production in the EU. In the analyzed period, they generated 43.1% of the overall agricultural output in the EU. Cyprus and Malta (deleted from the chart as outliers) are notable among new EU member states, featuring relatively high output intensity and land productivity. However, considering their share in the agricultural output of the EU, they cannot be deemed significant to European agriculture (0.21% of total share). The position of Slovenia implies slightly higher values of indicators used for determining the location of respective countries, but its share in the EU's agricultural output was only 0.3%. Countries that stand out from the analyzed group are Poland and Romania. From 2010 to 2019, they accounted for the highest percentage of agricultural output in the EU-13, that is, 5.7% and 4.1%, respectively. Hungary and Czechia had a similar position in terms of output intensity and land productivity, although they had considerably lower shares of the EU's output (1.9% and 1.2%, respectively).

Table 4. Land productivity in the member states of the European Union from 2010 to 2019 (EUR per 1 ha).

Country	2010	2019	2010–2019	Dynamics (2010 = 100)
Austria	1875.3	2482.1	2284	132.4
Belgium	5768.5	6464.8	6084.8	112.1
Bulgaria	690	724.1	697.3	104.9
Croatia	2135	1792.9	1883.4	84
Cyprus	5696.9	5992	5753.8	105.2
Czechia	1128.5	1286.5	1256.9	114
Denmark	3584.1	4477.2	3938.2	124.9
Estonia	647.3	926.8	780.8	143.2
Finland	1627.4	1687.2	1614.3	103.7
France	2257.3	2354.4	2326.5	104.3
Germany	2956.9	2953.4	2980.1	99.9
Greece	1874.6	2057.1	1959.5	109.7
Hungary	1121.9	1490.5	1343.6	132.9
Ireland	1274.3	1657.1	1456.1	130
Italy	3505.2	3318.6	3454.1	94.7
Latvia	488	676.8	586.1	138.7
Lithuania	699.9	883.8	834.3	126.3
Luxembourg	2389.9	2606.4	2517.7	109.1
Malta	10405.2	8910.2	9608.1	85.6
Netherlands	13,103.1	14,241.4	13,734.1	108.7
Poland	1343.3	1554.9	1464.5	115.8
Portugal	1752.5	1898.2	1809	108.3
Romania	995.5	1263.9	1115.8	127
Slovakia	916.5	1046.7	1051.8	114.2
Slovenia	2282.5	2481.7	2304.4	108.7
Spain	1653.3	2039.4	1831.8	123.4
Sweden	1622.4	1976.2	1769.4	121.8
United Kingdom	1310	1461.1	1390.2	111.5
EU-28	2682.3	2882.3	2779.7	107.5
Coefficient of variation (%)	108.3	101.3	104.4	-

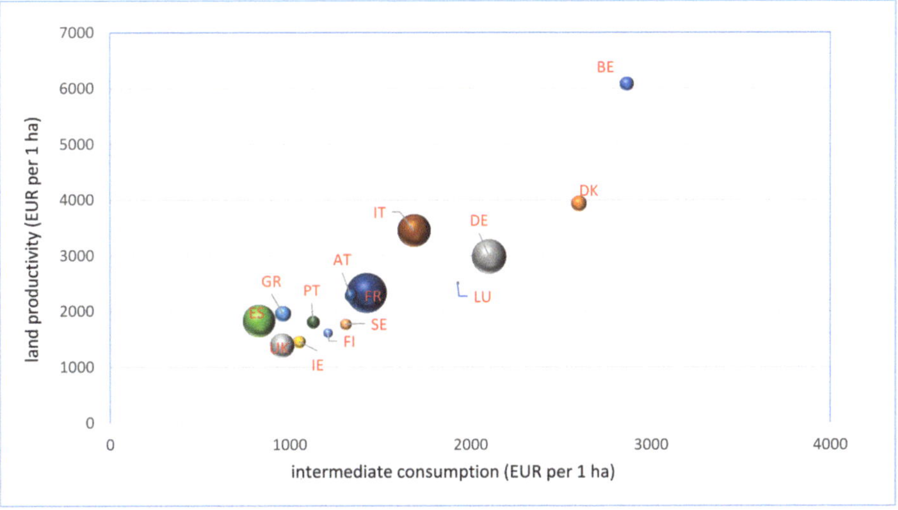

Figure 5. Position of countries forming the EU-15 depending on land productivity, intermediate consumption per 1 ha, and share in the value of agricultural production of the EU from 2010 to 2019. Abbreviations denoting EU-15 countries: Belgium, BE; Denmark, DK; Germany, DE; Ireland, IE; Greece, GR; Spain, ES; France, FR; Italy, IT; Luxembourg, LU; Netherlands, NL; Austria, AT; Portugal, PT; Finland, FI; Sweden, SE; United Kingdom, UK. Note: The size of the balls refers to the percentage of the EU's agricultural production.

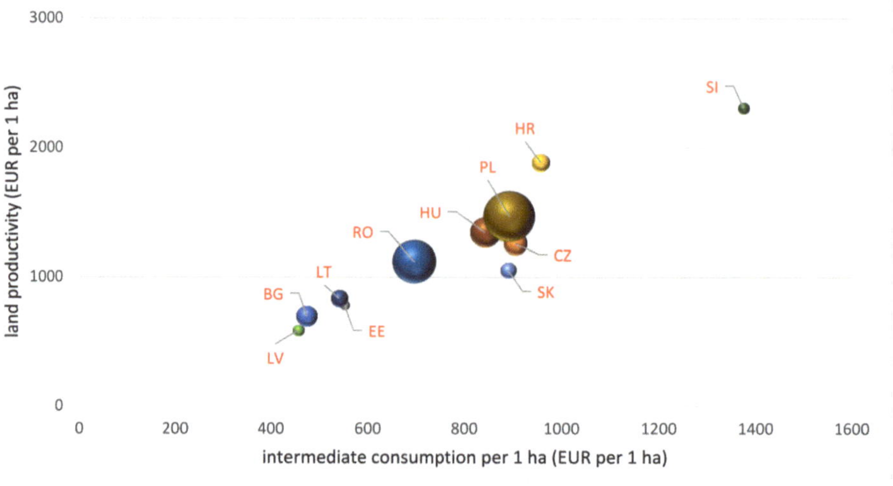

Figure 6. Position of countries forming the EU-13 depending on land productivity, intermediate consumption per 1 ha, and share in the value of agricultural production of the EU from 2010 to 2019. Abbreviations denoting EU-13 countries: Bulgaria, BG; Czechia, CZ; Estonia, EE; Croatia, HR; Cyprus, CY; Latvia, LV; Lithuania, LT; Hungary, HU; Malta, MT; Poland, PL; Romania, RO; Slovenia, SI; Slovakia, SK. Note: The size of the balls refers to the percentage of the EU's agricultural production.

Pawlak et al. [60] underlined that a long-term ability to maintain the high efficiency of agriculture affects its competitiveness and is also the basis for the transition from an industrial to a sustainable agriculture. According to the sustainable agriculture paradigm, natural resources can be used efficiently to achieve a satisfactory level of income from agricultural activity while respecting the laws of nature. However, it is difficult to specify the threshold of inputs that allow accomplishing sustainable development objectives. Ruiz-Martinez et al. [28] emphasized that few works define such thresholds using specific indicators. Some papers, such as those by Temme and Verburg [61], suggested future scenarios, including intensity thresholds based on nitrogen input. By contrast, Staniszewski [62] noted that, in countries such as Romania and Poland, increased resource productivity in agriculture is necessary for this sector's continuing growth and ensuring its competitiveness in the common EU market. At the same time, he pointed out that, in line with the concept of sustainable intensification in agriculture, this sector should achieve an increase in resource productivity without a detriment to the natural environment by implementing innovative production methods. Hunter et al. [34] highlighted that the objectives of sustainable intensification go beyond production output and performance to the extent of environmental protection. Additional political efforts are needed to manage the demand for food by reducing food wasting and changing eating habits [34,63,64].

Sustainable development is one of the greatest challenges to the world today, and accomplishing its objectives requires a compromise between economic growth and the environment [65]. Programs for sustainable development have been implemented for years; but, although the desired direction of change was set in agriculture, it is still insufficient for the perceived needs. At present, such opportunities should be sought in the European Green Deal that is expected to give rise to subsequent international measures to achieve ambitious climatic and environmental goals. However, the potential impact of that strategy on economic objectives remains a moot point. The ecological transformation postulated by the European Green Deal throws down challenges to countries, societies, agricultural producers, and institutions. These challenges refer to collaboration in research and production but also to undertaking measures to increase the social acceptance of environmental goals. We should also be aware that environmental protection and related environmental competitiveness are not only European but also global problems, implying a need for global solutions. Balancing the output intensity and environmental objectives requires an integrated approach at all levels of human activity.

4. Conclusions

Interest in the intensity of agricultural output stems from various challenges that agriculture needs to face. A major one is a need to orient it toward sustainable agriculture. Conventional agriculture was a product of the agriculture industrialization process, including output intensification, the concentration of the production potential, and the specialization of agricultural holdings. However, industrial agriculture's incontestable production and economic success were achieved at a considerable cost to the environment. This became a prerequisite for seeking new development directions based on the concept of sustainability. On these grounds, the diversification of agricultural output in the European Union member states was analyzed. This is an attempt at filling the research gap, which provides the basis for extended research considering a wider range of variables describing the production process in agriculture and its relationship with the natural environment. This paper has an interdisciplinary value as it combines economic and environmental aspects of agriculture that are relevant to sustainable development.

The results allowed us to formulate an answer to the abovementioned research questions concerning diversification of agricultural output intensity in the EU. Surveys showed that, from 2010 to 2019, member states of the European Union differed in terms of output intensity. However, these differences slightly decreased in the analyzed period, which is reflected in a reduced variability of costs per 1 ha of UAA. Most of the old EU member states from 2010 to 2019 were characterized by a higher output intensity than countries that joined the Community in or after 2004. In the analyzed period, the average value of intermediate consumption per 1 ha of utilized agricultural area in 13 new member states of the EU amounted to EUR 1275, while the mean for the EU-15 was EUR 2006.70. The land–labor ratio in the examined years ranged from 1.7 workers per 100 ha in the United Kingdom to 43 workers per 100 ha in Malta. Apart from Malta, countries with the highest number of workers per 100 ha were other new member states of the EU, particularly Slovenia, Cyprus, Croatia, Poland, and Romania. It was associated with the nature of agricultural structures in respective countries. Land productivity in EU member states was highly variable. The difference between the old and new member states was clearly marked. The highest land productivity from 2010 to 2019 was reached by agriculture in the Netherlands, where agricultural production levels per 1 ha were five times higher than on average in the European Union.

The surveys' outcomes imply a need for a diversified approach to sustainable agriculture within the EU, with a particular focus on differences between old and new member states. This applies to both the Common Agricultural Policy and national policies. Obviously, the challenges regarding changes in the development of agriculture should be different in countries where the levels of productivity and inputs are very high than in low-intensity ones. Considering the results of our research, one of the main challenges for the EU countries in the coming years should be pursuing a balance among economic, social, and environmental objectives in agricultural production.

Our research is not free of limitations and should be deemed preliminary. Measuring agricultural output intensity is subject to multiple issues regarding the methods due to the lack of unanimity in its evaluation. In addition, there are no limit thresholds for the analyzed indicators to allow evaluating agricultural output intensity. Thus, there is a need for continuing research using advanced econometric modeling techniques. This study will be aimed at designing a synthetic measure of sustainability of agriculture in the EU member states and assessing its dependence on agricultural output intensity.

Author Contributions: Conceptualization, A.N. and A.Z.; methodology, A.N. and A.Z.; formal analysis, A.N. and A.Z.; resources, A.N. and A.Z.; data curation, A.N. and A.Z.; writing—original draft preparation, A.N. and A.Z.; writing—review and editing, A.N. and A.Z.; visualization, A.N. and A.Z.; funding acquisition, A.N. and A.Z. All authors have read and agreed to the published version of the manuscript.

Funding: The APC was funded by the University of Life Sciences, Faculty of Agrobioengineering, and the vouchers. Publication was co-financed with the project entitled 'Excellent science' program of the Ministry of Education and Science as a part of the contract No. DNK/513265/2021 'Role of agriculture in implementing concept of sustainable food system "from field to table"'.

Institutional Review Board Statement: Not applicable.

Informed Consent Statement: Not applicable.

Data Availability Statement: Data used in surveys is sourced from EUROSTAT. Raw data were converted in accordance with the adopted methodology given in the publication. Data from databases are widely available for use. (https://ec.europa.eu/eurostat/data/database, accessed on 10 June 2022).

Conflicts of Interest: The authors declare no conflict of interests.

References

1. Czyżewski, A.; Czyżewski, B. Research challenges for agricultural economics in the new paradigm. In *Political Rents of European Farmers in the Sustainable Development Paradigm. International, National and Regional Perspective*; Czyżewski, B., Ed.; PWN: Warsaw, Poland, 2016; pp. 18–27.
2. Gołaś, M.; Sulewski, P.; Wąs, A.; Kłoczko-Gajewska, A.; Pogodzińska, K. On the way to sustainable agriculture—Eco-efficiency of Polish commercial farms. *Agriculture* **2020**, *10*, 438. [CrossRef]
3. European Commission. CAP Context Indicators 2014–2020. 33. Farming Intensity. Available online: https://ec.europa.eu/info/sites/default/files/food-farming-fisheries/farming/documents/2019-context-indicators-fiches.pdf (accessed on 10 March 2022).
4. Zegar, J.S. *Contemporary Challenges in Agriculture*; PWN: Warsaw, Poland, 2012. (In Polish)
5. Kremen, C.; Miles, A. Ecosystem services in biologically diversified versus conventional farming systems: Benefits, externalities, and trade-offs. *Ecol. Soc.* **2012**, *17*, 40. [CrossRef]
6. Żmija, D. Sustainable development of agriculture and rural areas in Poland. *Econ. Stud. Sci. J. Univ. Econ. Katow.* **2014**, *166*, 149–158. (In Polish)
7. Magrini, A. Correction to: Assessment of agricultural sustainability in European Union countries: A group-based multivariate trajectory approach. *AStA Adv. Stat. Anal.* **2022**. [CrossRef]
8. Mori, K.; Christodoulou, A. Review of sustainability indices and indicators: Towards a new City Sustainability Index (CSI). *Environ. Impact Assess. Rev.* **2012**, *32*, 94–106. [CrossRef]
9. Cataldo, R.; Crocetta, C.; Grassia, G.; Lauro, N.C.; Marino, M.; Voytsekhovska, V. Methodological PLS-PM Framework for SDGs System. *Soc Indic. Res.* **2021**, *156*, 701–723. [CrossRef]
10. Krasowicz, S. Relationships between man and natural environment in the aspect of sustainable development. *Probl. Inżynierii Rol.* **2008**, *1*, 21–27. (In Polish)
11. Radulescu, C.V.; Ioan, I. Sustainable development of Romanian agriculture within the context of European Union's requirements. *USV Ann. Econ. Public Adm.* **2015**, *15*, 57–62.
12. Oleszko-Kurzyna, B. Sustainable agriculture in the light of the European Union environmental requirements. *Probl. World Agric.* **2008**, *4*, 326–336.
13. Hayati, D.; Ranjbar, Z.; Karami, E. Measuring agricultural sustainability. In *Biodiversity, Biofuels, Agroforestry and Conservation Agriculture*; Sustainable Agriculture Reviews; Lichtfouse, E., Ed.; Springer: Dordrecht, The Netherlands, 2010; pp. 73–100.
14. Kelly, E.; Latruffe, L.; Desjeux, Y.; Ryan, M.; Uthes, S.; Diazabakana, A.; Dillon, E.; John Finn, J. Sustainability indicators for improved assessment of the effects of agricultural policy across the EU: Is FADN the answer? *Ecol. Indic.* **2018**, *89*, 903–911. [CrossRef]
15. Czyżewski, A.; Staniszewski, J. Dilemmas of operationalising the paradigm of sustainable agricultural development using the concept of eco-efficiency. *Probl. World Agric.* **2018**, *18*, 44–56. (In Polish) [CrossRef]
16. Bacon, C.; Getz, C.; Kraus, S.; Holland, K. The social dimensions of sustainability in diversified, industrial and hybrid farming systems. *Ecol. Soc.* **2012**, *17*, 41. [CrossRef]
17. Guth, M.; Smędzik-Ambroży, K.; Czyżewski, B.; Stępień, S. The Economic Sustainability of Farms under Common Agricultural Policy in the European Union Countries. *Agriculture* **2020**, *10*, 34. [CrossRef]
18. Pretty, J. Agricultural sustainability: Concepts, principles and evidence. *Philos. Trans. R. Soc. B: Biol. Sci.* **2008**, *363*, 447–465. [CrossRef]
19. Nowak, A.; Krukowski, A.; Różańska-Boczula, M. Assessment of sustainability in agriculture of the European Union countries. *Agronomy* **2019**, *9*, 890. [CrossRef]

20. Wrzaszcz, W.; Prandecki, K. Economic efficiency of sustainable agriculture. *Probl. Agric. Econ.* **2015**, *343*, 15–36. [CrossRef]
21. Komorowska, D. Development of modern agriculture in the context of sustainable development goals. *Village Agric.* **2014**, *3*, 71–84. (In Polish)
22. Keys, E.; McConnell, W.J. Global change and the intensification of agriculture in the tropics. *Glob. Environ. Change* **2005**, *15*, 320–337. [CrossRef]
23. Głowacki, M. Regional differentiation of agriculture intensity in Poland. *Pulawski Diary* **2002**, *130*, 213–221. (In Polish)
24. Kopiński, J. Tendencies of changes in agricultural production intensity in Poland in the aspect of potential environmental impacts. *Probl. World Agric.* **2011**, *11*, 95–104. (In Polish)
25. Teillard, F.; Allaire, G.; Cahuzac, E.; Léger, F.; Maigné, E.; Tichit, M. A novel method for mapping agricultural intensity reveals its spatial aggregation: Implications for conservation policies. *Agric. Ecosyst. Environ.* **2012**, *149*, 135–143. [CrossRef]
26. Barretto, A.; Berndes, G.; Sparovek, G.; Wirsenius, S. Agricultural intensification in Brazil and its effects on land-use patterns: An analysis of the 1975–2006 period. *Glob. Change Biol.* **2013**, *19*, 1804–1815. [CrossRef]
27. Levers, C.; Butsic, V.; Verburg, P.H.; Müller, D.; Kuemmerle, T. Drivers of changes in agricultural intensity in Europe. *Land Use Policy* **2016**, *58*, 380–393. [CrossRef]
28. Ruiz-Martinez, I.; Marraccini, E.; Debolini, M.; Bonari, E. Indicators of agricultural intensity and intensification: A review of the literature. *Ital. J. Agron.* **2015**, *10*, 74–84. [CrossRef]
29. Lambin, E.F.; Rounsevell, M.D.A.; Geist, H.J. Are agricultural land-use models able to predict changes in land-use intensity? *Agric. Ecosyst. Environ.* **2000**, *82*, 321–331. [CrossRef]
30. Sobczyński, T. Intensification and concentration of production and the economic and environmental sustainability of EU dairy and grain farms. *Ann. Pol. Assoc. Agric. Agribus. Econ.* **2011**, *13*, 154–159.
31. Foley, J.A.; DeFries, R.; Asner, G.P.; Barford, C.; Bonan, G.; Carpenter, S.R.; Chapin, F.S.; Coe, M.T.; Daily, G.C.; Gibbs, H.K.; et al. Global Consequences of Land Use. *Science* **2005**, *309*, 570–574. [CrossRef]
32. Bommarco, R.; Kleijn, D.; Potts, S.G. Ecological intensification: Harnessing ecosystem services for food security. *Trends Ecol. Evol.* **2013**, *28*, 230–238. [CrossRef]
33. Von Lampe, M.; Willenbockel, D.; Ahammad, H.; Blanc, E.; Cai, Y.; Calvin, K.; Fujimori, S.; Hasegawa, T.; Havlik, P.; Heyhoe, E.; et al. Why do global long-term scenarios for agriculture differ? An overview of the AgMIP Global Economic Model Intercomparison. *Agric. Econ.* **2014**, *45*, 3–20. [CrossRef]
34. Hunter, M.C.; Smith, R.G.; Schipanski, M.E.; Atwood, L.W.; Mortensen, D.A. Agriculture in 2050: Recalibrating targets for sustainable intensification. *Bioscience* **2017**, *67*, 386–391. [CrossRef]
35. Snapp, S.S.; Gentry, L.E.; Harwood, R. Management intensity—not biodiversity—the driver of ecosystem services in a long-term row crop experiment. *Agric. Ecosyst. Environ.* **2010**, *138*, 242–248. [CrossRef]
36. Steensland, A.; Zeigler, M. Productivity in agriculture for a sustainable future. In *The Innovation Revolution in Agriculture. A Roadmap to Value Creation*; Campos, H., Ed.; Springer: Cham, Switzerland, 2021; pp. 33–69.
37. Nowak, A.; Janulewicz, P.; Krukowski, A.; Bujanowicz-Haraś, B. Diversification of the level of agricultural development in the member states of the European Union. *Cah. Agric.* **2016**, *25*, 55004. [CrossRef]
38. Kijek, A.; Kijek, T.; Nowak, A.; Skrzypek, A. Productivity and its convergence in agriculture in new and old European Union member states. *Agric. Econ.* **2019**, *65*, 1–9. [CrossRef]
39. De Roest, K.; Ferrari, P.; Knickel, K. Specialisation and economies of scale or diversification and economies of scope? Assessing different agricultural development pathways. *J. Rural Stud.* **2018**, *59*, 222–231. [CrossRef]
40. Sadowski, A.; Wojtasiak, J. Production potential of agriculture in the countries of the European Union. *Zagadnienia Doradz. Rol.* **2019**, *1*, 5–19. (In Polish)
41. Pawłowska-Tyszko, J.; Osuch, D.; Płonka, R. *Standard 2020 Results Obtained by Farms Participating in the Polish FADN. Part I. Standard Results*; IERiGŻ-PIB: Warsaw, Poland, 2021. (In Polish)
42. Smędzik-Ambroży, K.; Rutkowska, M.; Kirbaş, H. Productivity of the Polish Agricultural Sector Compared to European Union Member States in 2004–2017 Based on FADN Farms. *Ann. Pol. Assoc. Agric. Aribus. Econ.* **2019**, *21*, 422–431. [CrossRef]
43. Baráth, L.; Fertő, I. Productivity and convergence in European agriculture. *J. Agric. Econ.* **2016**, *68*, 228–248. [CrossRef]
44. Ossowska, L.; Janiszewska, D. Employment and agricultural intensity of European Union countries. *Probl. World Agric.* **2018**, *18*, 238–247. [CrossRef]
45. Matyka, M. Trends in consumption of mineral fertilizers in poland against the background of the European Union. *Ann. Pol. Assoc. Agric. Aribus. Econ.* **2013**, *15*, 237–241.
46. Piwowar, A. Consumption of mineral fertilizers in the Polish agriculture trends and directions of changes. *Agric. Sci.* **2021**, *11*, 477–487. [CrossRef]
47. Hossain, M.; Singh, V.P. Fertilizer use in Asian agriculture: Implications for sustaining food security and the environment. *Nutr. Cycl. Agroecosyst.* **2000**, *57*, 155–169. [CrossRef]
48. Snyder, C.S.; Bruulsema, T.W.; Jensen, T.L.; Fixen, P.E. Review of greenhouse gas emissions from crop production systems and fertilizer management effects. *Agric. Ecosyst. Environ.* **2009**, *133*, 247–266. [CrossRef]
49. Frische, T.; Egerer, S.; Matezki, S.; Pickl, C.; Wogram, J. 5-Point programme for sustainable plant protection. *Environ. Sci. Eur.* **2018**, *30*, 8. [CrossRef]

50. Lawniczak, A.E.; Zbierska, J.; Nowak, B.; Achtenberg, K.; Grześkowiak, A.; Kanas, K. Impact of agriculture and land use on nitrate contamination in groundwater and running waters in central-west Poland. *Environ. Monit. Assess.* **2016**, *188*, 172. [CrossRef]
51. Zalewski, A. Changes in the Value of Used Fertilizers and Plant Protection Products in the Countries of the European Union in the Years 2010–2018. *Probl. World Agric.* **2020**, *20*, 78–87. [CrossRef]
52. Ciborowski, R. Technological innovations and the process of creating knowledge-based economy. In *Sustainable Development of Knowledge-Based Economy*; Poskrobko, B., Ed.; Higher School of Economics: Białystok, Poland, 2009; pp. 290–298.
53. Sadłowski, A.; Wrzaszcz, W.; Smedzik-Ambroży, K.; Matras-Bolibok, A.; Budzyńska, A.; Angowski, M.; Mann, S. Direct payments and sustainable agricultural development—The example of Poland. *Sustainability* **2021**, *13*, 13090. [CrossRef]
54. Smol, M. Transition to Circular Economy in the Fertilizer Sector—Analysis of Recommended Directions and End Users' Perception of Waste-Based Products in Poland. *Energies* **2021**, *14*, 4312. [CrossRef]
55. EUROSTAT. Agri-Environmental Indicator—Livestock Patterns. Available online: https://ec.europa.eu/eurostat/statistics-explained/index.php?title=Agri-environmental_indicator_-livestock_patterns#Livestock_density_at_EU_level_in_2016 (accessed on 18 August 2022).
56. Kopiński, J. Agri-environmental effects of changes in agricultural production in Poland. *Econ. Reg. Stud.* **2015**, *8*, 5–18.
57. Parzonko, A. Labour resources reserves in agriculture in Poland and possibilities of their use. *Ann. Pol. Assoc. Agric. Agribus. Econ.* **2016**, *18*, 292–297.
58. Tarnowska, A. Productivity of Chosen Production Factors in Agriculture in the European Union in the Years 2005–2012. *Ann. Pol. Assoc. Agric. Agribus. Econ.* **2014**, *16*, 214–219. (In Polish)
59. Zhang, L.; Hong, M.; Guo, X.; Qian, W. How Does Land Rental Affect Agricultural Labor Productivity? An Empirical Study in Rural China. *Land* **2022**, *11*, 653. [CrossRef]
60. Pawlak, K.; Smutka, L.; Kotyza, P. Agricultural potential of the EU Countries: How far are they from the USA? *Agriculture* **2021**, *11*, 282. [CrossRef]
61. Temme, A.J.A.M.; Verburg, P.H. Mapping and modelling of changes in agricultural intensity in Europe. *Agric. Ecosyst. Environ.* **2011**, *140*, 46–56. [CrossRef]
62. Staniszewski, J. Attempting to measure sustainable intensification of agriculture in countries of the European Union. *J. Environ. Prot. Ecol.* **2018**, *19*, 949–957.
63. West, P.C.; Gerber, J.S.; Peder, M.E.; Mueller, N.D.; Brauman, K.A.; Carlson, K.M.; Cassidy, E.S.; Johnston, M.; MacDonald, G.K.; Ray, D.K.; et al. Leverage points for improving global food security and the environment. *Science* **2014**, *345*, 325–327. [CrossRef]
64. Davis, K.F.; Gephart, J.A.; Emery, K.A.; Leach, A.M.; Galloway, J.N.; D'Odorico, P. Meeting future food demand with current agricultural resources. *Glob. Environ. Chang.* **2016**, *39*, 125–132. [CrossRef]
65. Góral, J.; Rembisz, W. Production in agriculture in the context of environmental protection. *J. Agric. Econ. Ext. Rural Dev.* **2017**, *104*, 7–21. [CrossRef]

agriculture

MDPI

Review

Approaches to Integrated Pest Management in Orchards: *Comstockaspis perniciosa* (Comstock) Case Study

Katarzyna Golan, Izabela Kot *, Katarzyna Kmieć and Edyta Górska-Drabik

Department of Plant Protection, University of Life Sciences in Lublin, Leszczyńskiego 7, 20-069 Lublin, Poland
* Correspondence: izabela.kot@up.lublin.pl

Abstract: Insect pests have major effects on agricultural production and food supply. Pest control in conventional crop management in orchards is mainly based on agrochemicals, which entails economic, health and environmental costs. Other approaches, such as biological methods or products based on biologically active substances and sex pheromones used for mating disruption, have faced some implementation challenges, particularly in relation to invasive insect species. The key for appropriate insect pest management is to identify the species and understand its biology and behaviour. Pest management systems should monitor, detect and inform fruit growers about changes in insect distribution, population ecology, possible damage level and economic loses. *Comstockaspis perniciosa* (San José scale—SJS) is a model example of a pest against which the method of integrated pest management should be adopted. This review presents a case study to support this statement.

Keywords: *Comstockaspis perniciosa*; SJS; monitoring; treatment decisions; pest management

1. Introduction

The armored scales (Hemiptera: Coccomorpha: Diaspididae) are sap-sucking, often concealed insects. There are approximately 2600 described species of Diaspididae worldwide [1], of which at least 200 are pests of agricultural crops, forest trees, ornamentals and greenhouse plants [2]. They are among the most invasive insects in the world, since they can be readily transported with plant material. Moreover, they are small and cryptic in habit, and thus go largely unnoticed and uncollected in natural habitats [3]. Armored scale insects are a morphologically distinct and homogenous group with extreme sexual dimorphism. Male nymphs have five instars, including two pupa-like quiescent stages, and adult males have a distinct head, thorax, an abdomen and one pair of wings. They live for a day or less and never feed [1]. Female nymphs have three instars, and the crawler instar (first instar) is the only one that is mobile. Adult females are morphologically reduced, sessile, legless, wingless and do not have a clear head and body [4]. The bodies of the older nymphs and females have a protective covering, the so-called 'scale', consisting of a waxy secretion, which is shaped with a pygidium. The scale-like covering either may occur as a discharge adhering to the cuticle or as a structure detached from the body [5]. This cover offers protection from direct contact with insecticides, which affects the effectiveness of chemical control. Furthermore, efforts to manage invasive insects, including scales, are often non-species-specific and provide short-lived effects [6]. The mouth parts of Diaspididae consist of piercing stylets that penetrate cells of mesodermal parenchyma in leaves and stems as well as mesocarp cells in fruits. Chlorosis, necrosis, reduced productivity, retarded development of the plants and their susceptibility to microbes and other insects is observed as a consequence of their feeding [7]. Because they feed on the parenchyma tissues of the host plants, they do not produce honeydew [8].

The control of the armored scale insects is based mainly on application of synthetic insecticides. The use of natural enemies and their efficacy is dependent on the correct identification of both the armored scale and its natural enemies, while the morphological identification of scale insects is based on microscopic cuticular characters of adult females, which requires expertise [9]. Parasitoids associated with Diaspididae belong to the families of Aphelinidae, Encyrtidae and Signiphoridae, with Aphelinidae being most numerously represented [10].

One of the armored scale insect species, *Comstockaspis perniciosa* (Comstock, 1881) (San José scale, SJS) previously known as *Quadraspidiotus perniciosus* (Comstock), fits in with the characteristics mentioned above. It is a cosmopolitan and highly polyphagous species, with a tendency towards rapid and mass colonisation of host plants. It is considered the most serious pest of fruit trees in Europe and several other countries in the world [11]. *C. perniciosa* is native to eastern Asia and was brought to North America, where it was first discovered in San José, California in 1880 [12]. Currently, it is common in the Palearctic and Nearctic regions, as well as in South America, Australia and New Zealand [13]. The San José scale had been the object of quarantine regulations in European countries; however, due to its spread over most of the continent, the European Commission decided not to qualify this species as a quarantine pest anymore [14]. Many abiotic and biotic factors affect the spread and evolution of *C. perniciosa*; nevertheless, temperature seems to be the most relevant [11].

The list of San José scale host plants includes at least 193 plant genera, with *Malus*, *Morus*, *Prunus*, *Pyrus*, *Ribes*, *Rubus* and *Vaccinium* being crucial due to their economic importance [1]. This species is described as a major pest, causing severe damage to almond and peach trees in Greece [15,16] and Ukraine [17], pears (destruction of trees) in Kazakhstan [18] and China [19], sour-cherry trees in Hungary [20] and apple trees in New Zealand [21], India [22], Romania [23], Kazakhstan [24], Chile [25], Portugal [26] and Poland [11].

The overwintering stage is the so-called black cap nymph (immature first instar), which occurs on the bark of tree trunks and branches. Hibernation of adult females or second-stage nymphs has sometimes been observed; however, only black cap nymphs were recorded to survive the low temperatures during the winter [11]. Development resumes in spring, and nymphs of *C. perniciosa* undergo several moults and the scales grow in diameter. The female's scale covering is circular, while the male's becomes elongated. The female's body is soft, yellow in colour and without wings or legs. Its covering is grey, reaching a diameter of 2 mm. Males are winged, 1 mm long and the body colour is yellow, with a dark band across the back. They fly for 2–3 days, mate with females and subsequently die. Females are viviparous and have a high fertility rate. First-instar nymphs, commonly referred to as crawlers (mobile stage), are oval, yellow and about 0.3 mm long. Only this stage is capable of dispersing and colonising new areas. They walk on bark, leaves or fruits until they find a suitable place to settle. They are active for less than one day, and factors such as temperature, humidity, dustiness, population density and host plant species have an impact on crawlers' settlement [27]. Within 24 h of hatching, they settle and insert their mouthpart into the host plant tissues, their antennae and legs undergo atrophy, they feed on sap and they start to produce cover. Feeding nymphs secrete a white waxy covering (white cap nymph), which turns black (black cap nymph) and then grey before maturation. According to the literature, two or three generations occur per year in the northern hemisphere [11,14,28,29] and three or four in the southern [25,30] hemisphere. Sometimes, single generations become so numerous that they overlap, and insects completely cover tree branches. Scale insects, including the San José scale, show low dispersal ability, and they spread by wind, birds, other insects or by infested seedlings in the nursery.

C. perniciosa, as all armored scales, feed on the content of individual parenchyma cells [31]. During feeding, sap accumulates in tree bark tissues, causing the surface to swell and the bark to crack. This results in a decrease in tree vigour, growth and productivity [32]. Long-lasting feeding, without pest control, can lead to the death of twigs or even the entire tree. Infested fruits have a slight depression and reddish-purple blotches around the feeding sites. This causes distortions, cracking and premature fruit dropping, ultimately reducing the quality and size of the yield (fruits are not marketable). Red spots may form around the scale within 24 h of crawler settling, but can also develop several weeks later. At present, the control of *C. perniciosa* is mainly based on the dormant application of mineral oils in wintertime against overwintering stages and the application of insecticides during the growing season. Summer treatments with mineral oils during the growing season are also possible due to improvements in refinement of oils, which are safer to plants. Nevertheless, there are still many precautions recommended (e.g., avoiding large spray droplet sizes by using the right equipment and spray pressure; oils must be sprayed directly on the insect due to their low residual activity, and oils cannot be sprayed when temperatures are below 5 °C or above 28 °C degrees and the relative humidity is above 90%) whenever using an oil on a woody plant to avoid plant injury (phytotoxicity) [33]. Furthermore, considering that the European Commission has proposed "Sustainable Use of Plant Protection Products" regulation with an objective to cut pesticides by 50% by 2030, and the fact that pesticide use against the San José scale is limited to period of crawler occurrence (the most sensitive stage to insecticide), this review updates key information on management practices [34]. Our approach was to review the current state of knowledge regarding monitoring and infestation assessment methods, the application of insecticides and nonchemical control methods to help develop control strategies of this pest.

2. Monitoring and Treatment Decisions

The presence of SJS is mostly detected on twigs and branches during pruning and on fruits during harvest or packing. Scouting the trees during the dormancy period allows one to detect infested plants and determine the level of infestation [14]. If the presence of SJS is detected, monitoring methods should be applied. Searching for crawlers and immobile instars on twigs, leaves and fruits, the application of sticky tape traps for crawlers, and pheromone traps for winged males are the most commonly used methods in SJS monitoring [11,14,35–37]. Assessing the number of SJS per fruit and the percentage of infested fruits allows one to estimate the degree of orchard infestation, and also provides feedback on the treatments applied. Results of the research conducted in Portugal demonstrated that 64.5% and 100% of fruits were infested in commercial and abandoned apple orchards, respectively [36].

Most San José scale specimens overwinter in the third phase of the first nymphal instar, known as the black cap stage. Winter survival is high, reaching more than 80% in untreated orchards [31]. This usually results in a well-synchronised emergence of the first generation of adults and crawlers in spring [38]. Adult males live for only a few days, while adult females produce offspring during a period of six weeks. In this way, successive generations overlap and all stages can occur on the tree at the same time during summer [30,31,38].

Research conducted in the 1970s and 1980s on SJS phenology and sex pheromones enabled the use of synthetic pheromones to detect and monitor male activity [39–44]. Since then, this method has been widely used by scientists and fruit growers around the world [14,45–48]. SJS sex pheromone compounds include (Z)-3,7-dimethy1-2,7-octadien-1-y1 propanoate and 7-methyl-3-methylene-7-octen-1-y1 propanoate. Synthetic sex pheromones have also been found to have a kairomonal effect on the SJS parasitoid *Encarsia perniciosi* (Hymenoptera: Aphelinidae) [49]. Observations of male activity under field conditions of North America and Greece indicated a lower flight temperature threshold of approx. 17 °C [26,42,45]. Pheromone lures, which are effective for 4 to 6 weeks, can be placed in delta (closed) traps, open-tent (sticky-board) traps or wing-shape traps [14]. Hoyt et al. [39] found tent traps to be more efficient than closed traps. However, the study by Rychla [29] suggested the comparable effectiveness of wing and delta traps, with the latter being more convenient to handle. Recently, there has been an increased interest in the use of digital sensors for pest monitoring [50,51]. A trap attached to a wireless networked digital camera (self-counting trap) can be used for scale insect monitoring, which has been proven for the California red scale, *Aonidiella aurantii* (Maskell) [37]. This type of trap for SJS catching is available on the market. It is a helpful time-saving tool for growers; however, more research is needed in this field. In commercial orchards, pheromone traps are commonly applied to establish the time of the first male capture, referred to as biofix [14,38]. This date is used for the accumulation of degree days. SJS traps should be placed in spring and located at a height of about 2 m in the northeastern part of the tree [39]. Badenes-Perez et al. [38] demonstrated that in Kern County, CA, USA, the relative density of trapped males was positively correlated with the crawler population density of the first generation. In Greece, Deligeorgidis et al. [46] did not find any relationship between captured adults and nymphs. On the other hand, a study of Mague and Reissig [52] in Wayne County, NY, USA, showed an inverse relationship between the cumulative pheromone trap catch of males and the total direct count of crawlers on trees. The discrepancies in the results could be related to different weather conditions between various US states and Europe. Other factors affecting observations may involve the number of generations developing each year and/or host species (almond, apple cv. Red Chief, apple cv. McIntosh, respectively).

San José scale mobile crawlers are deprived of waxy cover only within approximately 24 h of hatching [11]. Subsequently, they become sessile and start to produce a waxy sac, making the next larval instars and females less vulnerable to environmental stresses and insecticide treatments [53]. Therefore, knowledge of peak mobile crawler activity is fundamental for effective management. Their presence can be confirmed by searching for crawlers on branches, leaves and fruits [31]. Various studies reported different crawler densities depending on the estimation method used. Wearing and de Boer [30] found up to 31.6 crawler per cm^2 of bark, while approximately 100 nymphs per 100 cm^2 were detected by Mague and Reissig [54]. A three-minute count of crawlers on apple bark using a hand lens showed the presence of up to 500 individuals depending on generation and management model [31]. On the other hand, 8.5 to 65.2 nymphs per fruit were recorded in a commercial and unsprayed orchard in Portugal, respectively [36]. Monitoring crawlers in the field is problematic due to their size and difficulties in their identification; thus, it is also possible to use sticky tape traps to assess their abundance. It is known that armored scale crawlers mostly remain on the same plant on which they emerged. They move mainly vertically on the tree for several hours after emergence covering distances of up to 3 m, but often settle within 1 m of their sessile mother. Hence, sticky tapes placed around the branches allows one to evaluate crawler activity and density [27,36,38]. Double-sided sticky traps are recommended for use in commercial orchards as an effective and practical tool for monitoring SJS nymphs [55]. On the other hand, sticky-tape traps have been shown to be labour-intensive, and are therefore not widely selected for treatment decisions by growers and their consultants [38].

Research on insect phenology in correlation with temperature, referred to as growing degree days (DD), led to the development of insect models. They are a useful tool for predicting insect development and timing of treatments [56]. The degree-day accumulation method can be used to predict the appearance of subsequent developmental stages [14,56,57]. SJS phenology is not consistent across studies. The low temperature threshold in many studies was set at 10.5 °C [14,29,44,55,58] or 10.6 °C [31,37]; however, other values were also reported, e.g., 7.3 °C [28] and 10 °C [26,52,54]. Seasonal DD accumulation should begin when daily temperatures exceed the developmental threshold of 10.5 °C. It usually starts on January 1 (e.g., Arizona, southern Utah) or March 1 (e.g., western Colorado, Idaho, northern Utah) depending on temperature conditions [14]. For SJS the most important thing is the precise timing of the treatment, which controls the first generation of crawlers and can prevent fruit infestation. Since insect activity varies from year to year depending on weather conditions, calculating DD can help in scouting operations, e.g., setting traps or looking for crawlers. The baseline temperature and accumulation start date for calculating DD for the San José scale vary in the literature; therefore, it is difficult to compare the results calculated for different stages and generations (Table 1). In general, pheromone traps should be placed approximately at the pink stage of apple [11,14]; biofix (first male catch) was recorded at 84–140 DD, while the first crawlers were recorded at 196–294 DD. This indicates the need to develop the SJS model independently for different regions.

Table 1. Timing of San José scale 1st-generation events based on degree-day accumulations.

Biofix (DD)	Emergence of 1st-Generation Crawlers (DD)		Developmental Threshold (°C)	Accumulating DD Start Date	Country	Reference
	First	Peak				
94–140	360	510/550	10.0	1 March	USA (NY)	[55]
116	326 (210 after biofix);	-	10.0	1 March	Portugal	[26]
275	405 after biofix	600–700 after biofix	10.5	1 January/ 1 March	USA	[14]
Mid–late April	(196 after biofix)	(41–43 days after biofix)	10.5	After biofix	Northern Greece	[55]
285	534	-	10.5	1 March	India (Kashmir)	[59]
84	286–294	-	10.6	1 March	USA (Tennessee)	[31]
135	324 (189 after biofix)	-	10.6	-	Czech Republic	[29]

Biofix—first male catch from overwintering generation; DD—degree day.

3. Application of Insecticides

Of all phytophagous insects of apple orchards, *C. perniciosa* is a key pest in many commercial orchards almost all over the world. Without proper control, it causes tree death within a few years [25,59–69]. Effective control with chemicals is not satisfactory, due to its behaviour and differences in susceptibility between individual developmental stages [25,70,71].

Numerous attempts to control SJS resulted in the development of lime-sulphur spray, which was the first widely used insecticide spray in the United States, extensively applied for SJS control until 1922. Felt [72] documented that properly prepared and applied lime-sulphur applications gave satisfactory results in controlling SJS populations in orchards. In turn, the entire US apple industry was threatened in 1914 with extinction when lime-sulphur applications did not provide protection due to SJS resistance [73]. Petroleum oil has been used to reduce the abundance of various pest species since 1871 [74]. However, promising results regarding *C. perniciosa* control using lime sulphur have significantly slowed down research on the use of oils for SJS control. Ackerman [75], in his landmark study, proved that oil emulsions were more effective in SJS control compared to lime-sulphur. As a result, oils have become the dormant spray of choice in SJS control, and are also essential components in the control of this pest in current eradication programmes.

Until the late 1940s, the damage caused by SJS was very severe. However, with the introduction of long-lasting chlorinated hydrocarbon insecticides, namely DDT and other persistent insecticides, SJS nearly disappeared from crops [12,76]. DDT, as a foliar spray for *C. perniciosa* control, started to be widely used in orchards in 1945 [77]. DDT was first applied to control codling moth and quickly replaced the previously used lead arsenate or cryolite. Until then, annual applications of dormant or delayed dormant sprays of oil, lime sulphur or mixtures thereof were necessary for SJS control to prevent serious losses [75]. These treatments often caused damage to plants in orchards by exerting a phytotoxic effect. In fact, growers often omitted or postponed the application of dormant sprays until the necessity for control became apparent because of a significant increase in scale infestation [12,78,79]. The studies conducted in the late 1940s indicated that two or more DDT cover spray applications inhibited the growth of the San José scale population despite skipping dormant sprays during the past 3 to 5 years. In turn, a significant increase in SJS population was recorded in the fruit orchards that were not sprayed with DDT [77].

The 1950s was another period of significant increase in the occurrence of San José scale in fruit crops, particularly in North America. SJS were present in orchards that were repeatedly treated with synthetic organic insecticides, due to the development of resistance, as well as probably the disruption of the natural enemy complex [80]. Research on the control of this pest has been revived following the development of many organophosphorus insecticides. The fact that growers have used even more than eight chemical treatments for many years, mainly with organophosphate (OP) insecticides, has contributed to the harmfulness and invasiveness of this pest since the 1980s. However, the incidence of this pest has increased, indicating a major resurgence attributed to growers switching from dormant diesel spray to potent synthetic chemical pesticides. This strategy yielded excellent results in reducing San José scale incidence in the early years, but later, in recent years, it led to pest resurgence and a disruption of the natural enemy complex, and possibly the development of pesticide resistance [12,63,79,80]. In consequence, the SJS was one of the first documented cases of insect resistance to synthetic insecticide in the USA [25]. Currently, SJS control in various regions of the world involves different protection strategies based primarily on integrated pest management and country-specific regulations. Buzzetti et al. [63] and González [80] observed increased SJS infestation levels in Chilean orchards in the early 21st century. Acetylcholinesterase (AChE) inhibitors from the group of organophosphate (OP) insecticides, such as chlorpyrifos and methidathion, have been frequently used in many orchards to control this pest, and chemical control programmes have included 6–8 applications per season [81]. According to literature data [12,63,79,80,82], organophosphate insecticides have been the main alternative for pest control in apple orchards since their introduction to the Chilean market in the 1960s. At the beginning of the 21st century, new requirements of importing markets have forced the use of more selective insecticide alternatives [63]. This shift in management strategy gave excellent results in reducing the incidence of San José scale in the early years, but subsequently, it caused pest resurgence [65]. At the end of the 20th century, the use of broad-spectrum insecticides such as methyl parathion and chlorpyrifos, which had previously kept SJS incidence to a minimum, was abandoned or reduced [83]. New regulations have limited the use of such products in many countries, but new chemical compounds have not shown the same degree of control [84].

At present, in addition to OP insecticides, the pesticide market offers a variety of other types of agents, including a class of neuroactive insecticides from the neonicotinoid group (nicotinic receptor agonists), which act as an insect neurotoxin; a class of chemicals called sulphoximines (sulphoxaflor), which affect the central nervous system of insects; and a class of pesticides known as tetronic acid insecticides. Juvenile hormone analogues and insect growth regulators (e.g., pyriproxyfen and buprofezin) are used in SJS population control in many countries in the world. These products prevent larvae from developing into their adult stage, or they act as chitin synthesis inhibitors. Insect growth regulators (IGR) buprofezin and pyriproxyfen, or neurotoxic sulphoxaflor, have recently been registered in Chile and North America and can be used in San José scale control [85,86]. According to Michigan State University, Lorsban® (chlorpyriphos), Esteem® (pyriproxyfen) and Centaur® (buprofezin) are the most effective insecticides for early-season SJS control [87]. Foliar preparations of Lorsban may be used for dormant or delayed-dormant *C. perniciosa* control, either alone or in combination with oil. Esteem works as an IGR by inhibiting egg development, and the application of Esteem with oil controls the overwintering stages of SJS. Centaur is an IGR insecticide that acts on insect nymph stages by inhibiting chitin biosynthesis, thereby interfering with insect moulting. Centaur can be used in single applications, with oil as an additive or a penetrant surfactant for effective control. In recent years, spirotetramat, another IGR, has been developed, which is used alone or in combination with thiacloprid in commercial formulations; acetamiprid and thiacloprid are other examples of recently introduced agents applied against *C. perniciosa* [81,88–90].

Local sprays of mineral oils have been used for many years against various groups of insects, but most commonly to control scale insects in horticulture [14,31,91]. Products recommended for this purpose are the so-called horticultural oils (e.g., superior, supreme, or other similar weight of petroleum oil). Oils block insect spiracles, causing them to die from asphyxiation. They may also interact with insect fatty acids and interfere its metabolism [33]. High efficacy (93% mortality of San José scale nymph) was shown for 6% soybean oil and 3% petroleum oil applied in the dormant season [31]. Even a single application of the above concentrations reduced pest population for subsequent years. A similar effect can be achieved with a soybean oil concentration <3%, but only after two consecutive years of application. The application of these oils also reduces the effect of tree dieback observed in orchards not protected with oils. Similar results were obtained by Mesbah et al. [92] who applied heavy and light mineral oils. The highest reduction in the *C. perniciosa* population infesting pear trees (mortality >90%) was obtained for a light oil called Caple-2 (applied in the summer season), followed by the heavy oils Albolium oil®, Marsona oil® and Moxy oil® (applied during the winter season). Mineral oils for the control of scale insects are comparable in efficacy to chemical pesticides and even superior in terms of protection of natural enemies and the environment. The effectiveness of mineral oils is closely related to the timing of application. The San José scale was controlled by the normal orchard practice of dormant spraying with diesel–oil emulsion and by the complex of natural enemies, including the dominant aphelinid parasitoid (*Encarsia perniciosi* Tower) [78,93]. According to Alston and coauthors [14], the best approach against the wintering stage is to use delayed-dormant sprays. Horticultural oils are recommended, but insecticide should be added if the SJS infestation rate is high. Timing of horticultural oil application against the crawler stage is also important for effective management of this species [14,91]. Recently, an increase in SJS population has been observed in many European countries and in the United States, probably due to a general decline in the use of dormant oil sprays, partly due to their increasing costs [29,63,78,79,84]. However, the spectrum of plant protection products used to control SJSs varies depending on the growing region and regulations. It is also important that fruits with ecologically based pest management are widely introduced and preferred by consumers, which reduces the use of chemical agents and the number of treatments and gives priority to nonchemical control.

The widespread use of toxic chemicals to control scale insects has caused many problems, such as the unsatisfactory effectiveness of *C. perniciosa* management programmes, development of insect resistance to insecticides, environmental pollution and reduced populations of natural enemies. Alternative, effective and environmentally safe nonchemical methods are urgently needed.

4. Nonchemical Control

Botanical insecticides, often referred to as green pesticides, are a group of nonchemical agents that have been widely tested to control many pest species. There are several studies confirming the high efficacy of plant extracts against representatives of the family Diaspididae [94,95]. Fitiwy et al. [94] documented the effectiveness of an insecticide extracted from the seed kernels of neem tree (*Azadiractha indica* Jussieu) and tree tobacco (*Nicotinia glauca* Graham) in controlling the armored scale *A. aurantii*, a species related to *C. perniciosa*, feeding on orange trees. The high efficacy of azadirachtin against this insect species was also confirmed in other studies [95,96]. Although there has been no research on the use of essential oils directly on *C. perniciosa*, literature data indicate that some essential oils are highly effective against this group of insects. Formulations prepared from the essential oils of *Ambrosia maritima* L., *Origanum minutiflorum* O. Schwarz & P.H. Davis, *Cymbopogon nardus* (L.) Rendle and *Cymbopogon citratus* (DC.) Stapf. can be used as green insecticides against *Aulacaspis tubereularis* (News.) (Diaspididae). Among them, *O. minutiflorum* was the most effective, and caused more than 88% of insect mortality. Essential oil preparations affect scale insects both by contact and systemically. After spraying, the oil solution forms a barrier on the insect covers, which prevents their respiration. Residues of the essential oil solution can also penetrate plant tissues, be transported throughout the plant and consequently kill sucking insects [96].

The presence of the San José scale on fruits is a serious problem, not only for fruit growers and organic food producers, but also for their exporters. Attempts have been made to use ultrasound to eliminate *C. perniciosa* from the fruit surface, but the expected phytosanitary effect was not achieved. The insufficient effectiveness of this method was attributed to the specific morphological structure of the insect [97]. In contrast, good results were achieved by Endarto and Wicaksono [98] using high-pressure water (HPW with pressure 1000 psi). This method allowed them to destroy all stages of the pest present on the fruits. According to the authors, an additional preventive effect can be achieved by adding calcium polysulphide to the water, which, by changing the microclimate of the environment, discourages mobile nymphs (crawlers) from infesting the sprayed trees. This method is easy and low-cost, as well as safe for nontarget plants and arthropods, because it consists of washing the trees (mainly the stem part) using only water without pesticide. In the 1990s, an attempt was also made to remove *C. perniciosa* by fumigation [99]. Fumigation with methyl bromide (32 g/m^3) killed all infesting stages of this species on 'Red Delicious' apples in normal storage after 31 days and in controlled-atmosphere cold storage after 137 days. Total scale mortality on another apple variety ('Winesap') occurred after almost six months in both types of storage, if they had been previously fumigated. Moreover, the dosage required for 100% insect mortality can be detrimental to fruit quality [100]. Given the concern regarding fumigant residues, this type of method has not been implemented. The study of Chu [101] carried out in various storage options of fruits infested with SJS showed that temperature and atmosphere had discernible effects on the survival of these herbivores.

C. perniciosa populations are limited by natural enemies. There are many species of parasitoids and predators that are to a higher or lesser extent specialised against *C. perniciosa*. Data on the occurrence and role of these beneficial organisms come predominantly from India, as well as Pakistan, Greece and Romania, and they mainly include various species of ladybird (Coleoptera: Coccinnellidae) and chalcid wasps (Hymenoptera: Chalcidoidea).

Among predatory beetles, 10 species were recorded, 9 of which were ladybirds of the genus *Chilocorus*. The literature also indicates single species of ladybird from the following genera as natural enemies of the San José scale: *Coccinella*, *Lindorus*, *Oenopia*, *Pharoscymnus*, *Platynaspis* and *Sticholatis*; as well as a beetle from the family Cybocephalidae: *Cybocephalus fodori* Endrödy-Younga (Table 2).

There are not many concrete, quantitative data on the contribution of natural enemies to the control of San José scale in commercial orchards. According to Hix [102], their abundance in orchards is rather low. Therefore, repeated attempts have been made to introduce and colonise some species of natural enemies of *C. perniciosa*. An example of an introduced parasitoid is *Prospaltella perniciosi* Tower. This parasitoid was introduced in Greece in 1968 from France, and then was brought to the United States. Within 2 to 10 years, the parasitoid was found be well established, mainly in peach, apple and pear orchards. However, the level of pest parasitism was not satisfactory, and varied from 2 to 5% [103]. Similar results were obtained by the authors when conducting a study on a native parasitoid species, *Aphytis* spp. On the other hand, a much higher efficiency of natural SJS enemies was shown by Trandafirescu et al. [104], who used three predator species together with three parasitoid species and were able to reduce the population of the San José scale by more than 60% (Table 2). According to Khan [105], the release of 35 individuals of *Chilocorus infernalis* Mulsant per plant significantly reduced *C. perniciosa* infestation. As reported by Mesbah et al. [92], mineral oils may cause adverse effects on nontarget parasitoids. The latter authors reported that mineral oils (mostly Marsona oil®, Moxy oil® and CAPL-2) exhibited higher toxicity (16–30% mortality) against the San José scale parasitoid *Aphytis diaspidis* Howard.

Table 2. Predators and parasitoids recorded from *C. perniciosa* based on literature data.

	Species	Taxonomy	References
Predators	*Chilocerus infernalis* Mulsant 1853 syn *Chilocorus bijugus*, Mulsant 1856	Coleoptera, Coccinellidae	[59,105–110]
	Chilocerus bipustulatus (Linnaeus, 1758)	Coleoptera, Coccinellidae	[15,103,104,111]
	Chilocerus renipustulatus (L.G. Scriba, 1791)	Coleoptera, Coccinellidae	[104,111]
	Coccinella septempunctata (Linnaeus, 1758)	Coleoptera, Coccinellidae	[109,110]
	Exochomus quadripustulatus (Linnaeus, 1758)	Coleoptera, Coccinellidae	[104]
	Lindorus lophantae (Blaisdell, 1892)	Coleoptera, Coccinellidae	[103]
	Sticholotis marginalis, Kapur, 1956	Coleoptera, Coccinellidae	[109,110]
	Pharoscymnus fleksibilis (Mulsant, 1853)	Coleoptera, Coccinellidae	[109,110]
	Oenopia sauzeti Mulsant, 1866	Coleoptera, Coccinellidae	[110,112]
	Platynaspis saundersi (Crotch, 1874)	Coleoptera, Coccinellidae	[112]
	Cybocephalus fodori Endrody-Younga, 1965	Coleoptera, Cybocephalidae	[15,103]
Parasitoids	*Aphytis* spp.	Hymenoptera Chalcidoidea Aphelinidae	[103]
	Aphytis sp proclia group Walker	Hymenoptera Chalcidoidea Aphelinidae	[109–111]
	Aphytis diaspidis (Howard, 1881)	Hymenoptera Chalcidoidea Aphelinidae	[104,113]
	Aphytis maculicornis Masi, 1911	Hymenoptera Chalcidoidea Aphelinidae	[111]
	Aphytis mytilaspidis (Le Baron, 1870)	Hymenoptera Chalcidoidea Aphelinidae	[103,111]
	Azotus perspeciosus Girault,1916	Hymenoptera Chalcidoidea Aphelinidae	[109,110]
	Azotus kashmirensis Narayanan, 1961 *	Hymenoptera Chalcidoidea Aphelinidae	[35,110]
	Encarsia perniciosi (Tower, 1913) *syn. Prospaltella perniciosi*	Hymenoptera Chalcidoidea Aphelinidae	[104,109–111,113]
	Hispaniella lauri Mercet, 1911	Hymenoptera Chalcidoidea Aphelinidae	[114]
	Marietta carnesi (Howard, 1910) *	Hymenoptera Chalcidoidea Aphelinidae	[35]
	Teleterebratus perversus Compere & Zinna, 1955	Hymenoptera Chalcidoidea Encyrtidae	[108]

Table 2. *Cont.*

Species	Taxonomy	References
Holcotorax spp.	Hymenoptera Chalcidoidea Encyrtidae	[104]
Sympiesis spp.	Hymenoptera Chalcidoidea Eulophidae	[104]
Apantheles ssp	Hymenoptera Braconidae	[104]

* hyperparasitoid.

Microorganisms such as entomopathogenic fungi, bacteria, viruses and nematodes also reduce the number of insects. This particularly applies to entomopathogenic fungi. They occupy an important position among all biocontrol agents because of their route of pathogenicity, broad host range and ability to control, e.g., sap-sucking pests. It is important to emphasise the minimal negative effect of entomopathogenic fungi on nontarget organisms, for which reason they offer a safer alternative in IPM [115,116]. Buhroo [117] and Buhroo et al. [22] tested fungal pathogens against crawlers and nymphs of *C. perniciosa*. Three fungal species, *Beauveria bassiana* (Bals.), *Lecanicillium lucanii* (Zimmermann) Zare & Gams and *Metarhizium anisopliae* (Metschin.) showed the highest efficacy against the SJS. High mortality (>75%) was determined at a concentration of 15×10^5 conidia/mL on day 30 after treatment (10 days after the emergence of the first crawlers). *M. anisopliae* showed slightly lower efficacy against the SJS. However, entomopathogenic fungi should be used when complete eradication of the pest is not required, and some crop damage is acceptable. Therefore, entomopathogenic fungi should be applied in combination with other methods if the pest must be completely eliminated.

The San José scale has been traditionally controlled by pesticides and dormant oils; there are some additional biological methods used in the control of this insect pest. Its sex-pheromone is known and has been in use for decades as a tool for monitoring *C. perniciosa*. However, increased pressure of SJS during recent years is providing a reason to look at new way of using pheromone for reducing this pest abundance. A new management strategy is mating disruption (MD) as a method for their control. MD is based on the release of synthetic sex pheromones, aiming to interrupt mate-finding communication and prevent mating in the target pest [68,84]. Males of San José scale are weak flyers; they only fly for a very short distance, while females are wingless, and this feature makes this species a very good species for testing pheromone-mediated mating disruption as an alternative strategy to insecticides [84]. Mating disruption has been commercially developed and applied against the vine mealybug *Planococcus ficus* (Signoret) (Hemiptera: Pseudococcidae) and the California red scale *A. aurantii*. Critical factors affecting MD effectiveness are pest density and effective disruption late in the season. According to literature data [68,69] MD applied to scale insect pests is more effective in small plots and compatible with biological control and integrated management programs. However, there are no commercially registered San José scale disruption products; research on this is still ongoing. The key factors for its commercial application are technological advances in pheromone synthesis and pheromone formulations. According to research conducted by Maas [69], Franco et al. [68] and Gut [84] in recent years, the potential for mating disruption as a pest control of San José scale seems high.

5. Concluding Remarks and Perspectives

The problems associated with SJS control are influenced by changes in plant protection programmes applied in Europe and worldwide, which recommend limiting the use of chemical plant protection products to the minimum necessary. Currently recommended insecticides are highly selective for the pest; hence, growers require a sustainable chemical control strategy for *C. perniciosa* based on accurate data of its biology and behaviour. The best strategy for managing San José scale is to prevent serious infestations, and the most optimal cultural control is to prune out infested branches. This reduces the number of scales and opens up tree canopies, allowing better spray penetration. It is therefore necessary to develop and promote precise SJS monitoring systems using new technologies. Further research is required on new solutions of using products based on biologically active substances and environmentally friendly pest management tactics for mating disruption based on the release of synthetic sex pheromones.

Author Contributions: Conceptualization, K.G. and K.K.; methodology, K.G., I.K., K.K. and E.G.-D.; formal analysis, K.G.; investigation, K.K.; resources, E.G.-D. and I.K.; writing—original draft preparation, K.G., I.K., K.K. and E.G.-D.; writing—review and editing, I.K.; supervision, K.G. All authors have read and agreed to the published version of the manuscript.

Funding: This research was funded by the University of Life Sciences in Lublin, grant number OKK/S/44, in 2019–2022.

Institutional Review Board Statement: Not applicable.

Data Availability Statement: All the required data relevant to the presented study are included in the manuscript.

Conflicts of Interest: The authors declare no conflict of interest.

References

1. García Morales, M.; Denno, B.D.; Miller, D.R.; Miller, G.L.; Ben-Dov, Y.; Hardy, N.B. ScaleNet: A literature-based model of scale insect biology and systematics. *Database* **2016**, bav118. [CrossRef]
2. Moran, P.J.; Goolsby, A.A. Biology of the armored scale *Rhizaspidiotus donacis* (Hemiptera: Diaspididae), a candidate agent for biological control of giant reed. *Ann. Entomol. Soc. Am.* **2010**, *103*, 252–263. [CrossRef]
3. Normark, B.B.; Morse, G.E.; Krewinski, A.; Okusu, A. Armored scale insects (Hemiptera: Diaspididae) of San Lorenzo National Park, Panama, with description of two new species. *Ann. Entomol. Soc. Am.* **2014**, *107*, 37–49. [CrossRef]
4. Normark, B.B.; Okusu, A.; Morse, G.E.; Peterson, D.A.; Itioka, T.; Schneider, S.A. Phylogeny and classification of armored scale insects (Hemiptera: Coccomorpha: Diaspididae). *Zootaxa* **2019**, *4616*, 001–098. [CrossRef] [PubMed]
5. Gullan, P.J.; Cook, L.G. Phylogeny and higher classification of the scale insects (Hemiptera: Sternorrhyncha: Coccoidea). *Zootaxa* **2007**, *1668*, 413–425. [CrossRef]
6. McLaughlin, G.M.; Dearden, P.K. Invasive insects: Management methods explored. *J. Insect Sci.* **2019**, *19*, 17. [CrossRef]
7. Sabree, Z.L.; Huang, C.Y.; Okusu, A.; Moran, N.A.; Normark, B.B. The nutrient supplying capabilities of *Uzinura*, an endosymbiont of armoured scale insects. *Environ. Microbiol.* **2013**, *15*, 1988–1999. [CrossRef]
8. Mansour, R.; Grissa-Lebdi, K.; Mazzeo, G.; Russo, A. Key scale insects (Hemipetar: Coccoidea) of high economic importance in Mediterranean area: Host plants, bio-ecological chcracteristics, natural enemies and pest management strategies—A review. *Plant Prot. Sci.* **2017**, *53*, 1–14. [CrossRef]
9. Amouroux, P.; Crochard, D.; Correa, M.; Groussier, G.; Kreiter, P.; Roman, C.; Guerrieri, E.; Garonna, A.; Malausa, T.; Zaviezo, T. Natural enemies of armored scales (Hemiptera: Diaspididae) and soft scales (Hemiptera: Coccidae) in Chile: Molecular and morphological identification. *PLoS ONE* **2019**, *14*, e0205475. [CrossRef]
10. Rehmat, T.; Anis, S.B.; Khan, M.T.; Fatma, J.; Begum, S. Aphelinid parasitoids (Hymenoptera: Chalcidoidea) of armoured scale insects (Homoptera: Diaspididae) from India. *Biol. Med.* **2011**, *3*, 270–281.
11. Golan, K. Contribution to the knowledge of the San José scale (Hemiptera, Coccomorpha, Diaspididae) in Poland. *Polish J. Entomol.* **2020**, *89*, 7–19. [CrossRef]
12. Blatchford, B.R. Pest panic in the American West: The San Jose scale as change agent in American agriculture, 1880–1900. Ph.D. Thesis, University of British Columbia, Vancouver, BC, Canada, 2017. [CrossRef]
13. Kozar, F. Forecasting and Monitoring Infestations. Forecasting. In *Armored Scale Insects their Biology, Natural Enemies and Control. Vol. B.*; Rosen, D., Ed.; Elsevier Science Publishers B.V.: Amsterdam, The Netherlands, 1990; pp. 335–340.
14. Alston, D.G.; Murray, M.; Reding, M.E. San Jose Scale (*Quadraspidiotus perniciosus*). *Utah Pests Fact Sheet* **2011**, *153*, 1–5.
15. Katsoyannos, P.I.; Argyriou, L. The phenology of the San Jose scale *Quadraspidiotus perniciosus* [Hom.: Diaspididae] and its association with its natural enemies on almond trees in northern Greece. *Entomophaga* **1985**, *30*, 3–11. [CrossRef]
16. Paloukis, S.S.; Navrozidis, E.I. Effectiveness of a new insecticide (Diofenolan) for control of San Jose scale, *Quadraspidiotus perniciosus* (Comstock) (Diaspididae), on peach trees in northern Greece. *Isr. J. Entomol.* **1995**, *29*, 285–286.
17. Yudytska, I.; Klechkovskyi, Y. Species composition of harmful entomocomplex in peach orchards of southern Ukraine. *Sci. Horiz.* **2021**, *24*, 61–67. [CrossRef]

18. Folkina, N.Y. A case of mass extermination of the San José scale (Quadraspisiotus perniciosus) by the ant *Creamtogaster subdentata. Zool. Zhurnal* **1978**, *5*, 301.
19. Jian, H.; Xiaonan, L. New record of fruit tree pest of Fujian—San Jose scale *Quadraspidiotus perniciosus* (Comstock) and its parasitoids. *J. Fujian Agric. Univ.* **1996**, *25*, 334–338.
20. Balázs, K.; Jenser, G. Significance of the parasitoids and predators in IPM of sour-cherry. *Integr. Plant Prot. Stone Fruit IOBC/Wprs Bull.* **2004**, *27*, 3–7.
21. Richards, A.M. Damage to apple crops infested with San José scale, *Quadraspidiotus perniciosus* (Comst.). *N. Z. J. Agric. Res.* **1962**, *5*, 479–484. [CrossRef]
22. Buhroo, A.A.; Rasheed, F.N.; Khanday, A.L. An assessment on population density of San Jose Scale *Quadraspidiotus perniciosus* (Comstock) and its biological control in Kashmir (Hemiptera: Diaspididae). *Mun. Entomol. Zool.* **2016**, *11*, 434–440.
23. Beşleaga, R.; Cârdei, E.; Georgescu, T.; Tălmaciu, M.; Corneanu, G. Control of San José scale (*Quadraspidiotus perniciosus* Comst.) from apple tree plantation at the fruit growing research and development station of Iaşi. *Cercet. Agron. Mold.* **2009**, *42*, 27–32.
24. Kopzhassarov, B.; Beknazarova, Z. On the problem of studying the features of development of San Jose scale (*Quadraspidiotus perniciosus* Comst.) in orchards in the south-east Kazakhstan. *News Natl. Acad. Sci. Repub. Kazakhstan. Ser. Agric. Sci.* **2016**, *1*, 30–35.
25. Buzzetti, K.; Chorbadjian, R.A.; Nauen, R. Resistance management for San Jose scale (Hemiptera: Diaspididae). *J. Econ. Entomol.* **2015**, *108*, 2743–2752. [CrossRef] [PubMed]
26. Rodrigues, A.N.; Torres, L.M.; Polesny, F. Phenology of San José scale, *Quadraspidiotus perniciosus* (Comstock) on apple in Guarda region (central eastern Portugal). *IOBC/WPRS Bull.* **2001**, *24*, 195–199.
27. Magsig-Castillo, J.; Morse, J.G.; Walker, G.P.; Bi, J.L.; Rugman-Jones, P.F.; Stouthamer, R. Phoretic dispersal of armored scale crawlers (Hemiptera: Diaspididae). *J. Econ. Entomol.* **2010**, *103*, 1172–1179. [CrossRef]
28. Bajec, D.; Knapič, M.; Rodič, K.; Brence, A.; Knapič, V.; Peterlin, A.; Zajc, M.; Vrtin, D. *Poročilo o Prerazmnožitvi Ameriškega Kaparja (Diaspidiotus perniciosus [Comst.], Sin.* Quadraspidiotus perniciosus *[Comst.]);* JV Sloveniji. KGZS–Zavod NM: Novo Mesto, Slovenia, 2010; p. 41.
29. Rychla, K. Monitoring of San José Scale (*Diaspidiotus perniciosus*) Occurrence and Comparison of Temperature Models. 2014. Available online: https://mnet.mendelu.cz/mendelnet2014/articles/50_rychla_1043.pdf (accessed on 15 November 2022).
30. Wearing, C.H.; de Boer, J.A. Temporal distribution of San José scale *Diaspidiotus perniciosus* (Hemiptera:Diaspididae) on an apple tree. *N. Z. Entomol.* **2014**, *37*, 61–74. [CrossRef]
31. Hix, R.L.; Pless, C.D.; Deyton, D.E.; Sams, C.E. Management of San Jose scale on apple with soybean-oil dormant sprays. *Hortscience* **1999**, *34*, 106–108. [CrossRef]
32. Charles, J.C.; Henderson, R.C. Catalogue of the exotic armoured scale insects (Hemiptera: Coccoidea: Diaspididae) in New Zealand. *J. R. Soc. N. Z.* **2002**, *32*, 587–615. [CrossRef]
33. Cranshaw, W.S.; Baxendale, B. Insect control: Horticultural oils. *Fact Sheet-Insect Ser.* **2013**, *5*, 1–3.
34. Available online: https://ibma-global.org/wp-content/uploads/2022/06/2022-06-22-SUR-Proposal-pesticides_sud_eval_2022_reg_2022-305_en-2.pdf (accessed on 26 October 2022).
35. Amin, M.M.; Trail, A.R. Seasonal history and biological control of San Jose Scale, *Quadraspidiotus perniciosus* (Diaspididae: Homoptera) on apple in Kashme. *J. Biol. Control* **1987**, *1*, 3–6.
36. Torres, L.M.; Rodrigues, A.N.; Avilla, J. Chemical control of *Quadraspidiotus perniciosus* (Comstock) (Homoptera: Diaspididae) in apples and side effects on phytoseiid mites (Acari: Phytoseiidae). *Integr. Fruit Prod./Wprs Bull.* **2001**, *24*, 207–212.
37. Frewin, A.; Lopez, B.; Cox, A.; Hoffman, E.; Hazell, J. Comparison of two traps for monitoring California red scale (Hemiptera: Diaspididae). *Fla Entomol.* **2019**, *102*, 586–591. [CrossRef]
38. Badenes-Perez, F.R.; Zalom, F.G.; Bentley, W.J. Are San José scale (Homoptera: Diaspididae) pheromone trap captures predictive of crawler densities? *J. Appl. Entomol.* **2002**, *126*, 545–549. [CrossRef]
39. Hoyt, S.C.; Westigard, P.H.; Rice, R.E. Development of pheromone trapping techniques for male San Jose scale (Homoptera: Diaspididae). *Environ. Entomol.* **1983**, *12*, 371–375. [CrossRef]
40. Rice, R.E. San Jose scale: Field studies with a sex pheromone. *J. Econ. Entomol.* **1974**, *67*, 561–562. [CrossRef]
41. Gieselmann, M.J.; Rice, R.E.; Jones, R.A.; Roelofs, W.L. Sex pheromone of the San José scale. *J. Chem. Ecol.* **1979**, *5*, 891–900. [CrossRef]
42. Rice, R.E.; Hoyt, S.C. Response of San Jose scale to natural and synthetic sex pheromones. *Environ. Entomol.* **1980**, *9*, 190–194. [CrossRef]
43. Rice, R.E.; Jones, R.A. Monitoring flight patterns of male San Jose scale (Homoptera: Diaspididae). *Can. Entomol.* **1977**, *109*, 1403–1404. [CrossRef]
44. Rice, R.E.; Flaherty, D.L.; Jones, R.A. Monitoring and modeling San Jose scale. *Calif. Agricul.* **1982**, *36*, 13–14.
45. Kyparissoudas, D.S. The occurrence of *Encarsia perniciosi* in areas of Northern Greece as assessed by sex pheromone traps of its host *Quadraspidiotus pernicisus. Entomol. Hell.* **1987**, *5*, 7–12. [CrossRef]
46. Deligeorgidis, P.N.; Kayoglou, S.; Sidiropoulus, M.; Variopolou, D.G.; Greveniotis, V.; Ipsilandis, C.G. Monitoring and control of *Quadraspidiotus perniciosus* Comstock (Hemiptera: Diaspididae) on apple trees in the prefecture of florin, greece. *J. Entomol.* **2008**, *5*, 381–388. [CrossRef]

47. Lo, P.L.; Wallis, R.; Bellamy, D.E. The effectiveness of two types of adhesive for catching insects in traps. *N. Z. Plant Prot.* **2019**, *72*, 230–236. [CrossRef]
48. Rozova, L.; Yudytska, I. Entomocomplex of peach plantations in the vegetation period. *Quar. Plant Prot.* **2020**, *10–12*, 24–26. [CrossRef]
49. Bayoumy, M.H.; Kaydan, M.B.; Kozar, F. Are synthetic pheromone captures predictive of parasitoid densities as a kairomonal attracted tool? *J. Entomol. Acarol. Res.* **2011**, *43*, 23–31. [CrossRef]
50. Potamitis, I.; Eliopoulos, P.; Rigakis, I. Automated Remote Insect Surveillance at a Global Scale and the Internet of Things. *Robotics* **2017**, *6*, 19. [CrossRef]
51. Zhu, S.; Malmqvist, E.; Li, W. Insect abundance over Chinese rice fields in relation to environmental parameters, studied with a polarization-sensitive CW near-IR lidar system. *Appl. Phys. B* **2017**, *123*, 211. [CrossRef]
52. Mague, D.L.; Reissig, W.H. Phenology of the San José scale (Homoptera: Diaspididae) in New York State apple orchards. *Can. Entomol.* **1983**, *115*, 717–722. [CrossRef]
53. Robayo, E.; Chong, J.-H. General biology and current management approaches of soft scale pests (Hemiptera: Coccidae). *J. Integ. Pest Mngmt.* **2015**, *6*, 17. [CrossRef]
54. Mague, D.L.; Reissig, W.H. Airborne dispersal of San José scale, *Quadraspidiotus perniciosus* (Comstock) (Homoptera: Diaspididae), crawlers infesting apple. *Environ. Entomol.* **1983**, *12*, 692–696. [CrossRef]
55. Kyparissoudas, D.S. Determination of spray dates for the control of the first generation of *Quadraspidiotus perniciosus* in Northern Greece. *Entomol. Hell.* **1990**, *8*, 5–9. [CrossRef]
56. Murray, M.S. Using Degree Days to Time Treatments for Insect Pests. *Utah Pests, Fact Sheet. IPM General, All Current Publications*. 2020, p. 978. Available online: https://core.ac.uk/download/pdf/77520811.pdf (accessed on 12 November 2022).
57. Vafaie, E.; Merchant, M.; Xiaoya, C.; Hopkins, J.D.; Robbins, J.A.; Chen, Y.; Gu, M. Seasonal population patterns of a new scale pest, *Acanthococcus langerstroemiae* Kuwana (Hemiptera: Sternorrhyncha: Eriococcidae), of Crapemyrtles in Texas, Louisiana, and Arkansas. *J. Environ. Hort.* **2020**, *38*, 8–14.
58. Jorgensen, C.D.; Rice, R.E.; Hoyt, S.C.; Westigard, P.W. Phenology of the San José scale (Homoptera: Diaspididae). *Can. Entomol.* **1981**, *113*, 149–159. [CrossRef]
59. Rahman, M.H.; Ghani, M.A.; Kazimi, S.K. Introduction of exotic natural enemies of San José scale in Pakistan. *Tech. Bull. Commonw. Inst. Biol. Control.* **1961**, *1*, 165–182.
60. Westigard, P.H.; Calvin, L. D1977. Sampling San José in a pest management program on pear in Southern Oregon. *J. Econ. Entomol.* **1977**, *70*, 138–140. [CrossRef]
61. Rosen, D. *Armored Scale Insects: Their Biology, Natural Enemies and Control*; Elsevier Science: Amsterdam, The Netherlands, 1990; Volume 4A.
62. Simeria, G. Cercetari cu privire la biologia si combaterea paduchelui testos din San Jose (*Quadraspidiotus perniciosus* Comst.) ın SV tarii (Investigations concerning San Jose scale (*Quadraspidiotus perniciosus* Comst.) biology and control. *Lucr. Stiintifice* **2001**, *20*, 226–229.
63. Buzzetti, K.A.; Chorbadjian, R.A.; Fuentes-Contreras, E.; Gutiérrez, M.; Ríos, J.C.; Nauen, R. Monitoring and mechanisms of organophosphate resistance in San Jose scale, Diaspidiotus perniciosus (Hemiptera: Diaspididae). *J. Appl. Entomol.* **2015**, *140*, 507–516. [CrossRef]
64. Masoodi, M.A.; Bhagat, C.K.; Sofi, R.M. Toxicity of insecticides to Encarsia perniciosi Tower and Aphytis proclia Walker. *J. Biol. Cont.* **1993**, *7*, 37–39.
65. Mohit, H.; Jagdeesh, P.R.; Anil, S.; Azeem, R.; Injila, Q.; Abdul, W.W. Description and management strategies of important pests of pear: A review. *J. Entomol. Zool. Stud.* **2018**, *6*, 677–683.
66. Zaki, F.A.; Mantoo, M.A.; Parray, M.A.; Khan, A.A.; Rather, A.Q.; Wani, N.A. Paraffinic mineral oils—A new ecofriendly agrochemical strategy for agricultural pest management. In Proceedings of the IUPAC Sponsored First International Conference On Agrochemicals Protecting Crop, Health And Natural Environment, New Delhi, India, 8–11 January 2008.
67. Nissar, T.; Gull, A.; Mir, M.A. Infestation rate of San Jose scale, *Quadraspidiotus perniciosus* (Comstock) in the fruit orchards of Baramulla, Kashmir, India. *J. Entomol. Zool. Stud.* **2020**, *8*, 928–932.
68. Franco, J.C.; Cocco, A.; Lucchi, A.; Mendel, Z.; Suma, P.; Vacas, S.; Mansour, R.; Navarro-Llopis, V. Scientific and technological developments in mating disruption of scale insects. *Entomol. Generalis* **2022**, *42*, 251–273. [CrossRef]
69. Maas, J. San Jose Scale Mating Disruption in Apples. Master's Thesis, Michigan State University, East Lansing, MI, USA, 2022.
70. Stoetzel, M.B. Seasonal history of seven species of armored scale insects of the *Aspidiotini* (Homoptera: Diaspididae). *Ann. Entomol. Soc. Am.* **1975**, *68*, 489–492. [CrossRef]
71. González, R. *Biología, Ecología y Control de la Escama de San José en Chile:* Quadraspidiotus perniciosus *(Comst)*; Universidad de Chile, Facultad de Ciencias Agronómicas: Santiago, Chile, 1981; Volume 9, pp. 1–64.
72. Felt, E.P. *Report of the State Entomologist*; New York State Museum: Albany, NY, USA, 1907; Volume 124, p. 38.
73. Melander, A.L. Varying susceptibility of the San Jose scale to sprays. *J. Econ. Entomol.* **1914**, *7*, 167–173. [CrossRef]
74. Riley, C.V. The kerosene emulsion: Its origin, nature, and increasing usefulness. *Proc. Annu. Meet. Soc. Promot. Agric. Sci.* **1892**, *12–13*, 83–98.
75. Ackerman, A.J. *Preliminary Report on Control of San Jose Scale with Lubricating—Oil Emulsion*; U.S. Department of Agriculture: Washington, DC, USA, 1923; p. 263.

76. Reissig, W.H.; Weires, R.W.; Onstad, D.W.; Stanley, B.H.; Stanley, D.M. Timing and effectiveness of insecticide treatments against the San Jose Scale (Homoptera: Diaspididae). *J. Econ. Entomol.* **1985**, *78*, 238–248. [CrossRef]
77. O'Neill, W.J. DDT and Parathion for SanJose Scale control. *J. Econ. Entomol.* **1951**, *44*, 711.
78. Sofi, M.A. Studies on the Current Status of San Jose Scale, *Quadraspidiotus Perniciosus* (Comstock) and Its Management on Apple. Ph.D. Thesis, Sher-e-Kashmir University of Agriculture Science and Technology of Kashmir, Srinagar, India, 2006; p. 158.
79. Sofi, M.A.; Khan, Z.H. Management of san jose scale, *Quadraspidiotus perniciosus* Comstock in apple orchards of Kashmir through horticultural mineral oils. *J. Entomol. Res.* **2021**, *16*, 32–35. [CrossRef]
80. Zaki, F.A.; Lone, A.H. Resistance in European red mite (Panonychus ulmi Koch.) to some pesticides on apple in Kashmir. In Proceedings of the International Symposium on Temperate Zone Fruits in the Tropics and Subtropics Sponsored by International Society for Horticultural Sciences, Solan, India, 14–18 October 2003; pp. 14–18.
81. Gonzalez, R.H. *Seminario Identificacio'n, Biologı'a y Manejo Fitosanitario de los Insectos Coccoideos de Importancia en Fruticultura y Viticultura (Escamas, Conchuelas, Chanchitos Blancos, Margarodes de la vid)*; Universidad de Chile, Facultad de Ciencias Agrono´micas, Direccio´n de Extensio´n: Santiago, Chile, 2012; p. 142.
82. Gonzalez, R.H. *Seminario Escama de San Jose Diaspidiotus perniciosus (Comstock) Plaga de Frutales de Hoja Caduca*; Universidad de Chile, Facultad de Ciencias Agronomicas, Departamento de Sanidad Vegetal: Santiago, Chile, 2011.
83. Food Quality Protection Act 1996. Public Law 104–170. Available online: https://www.govinfo.gov/app/details/PLAW-104publ170 (accessed on 25 October 2022).
84. Available online: https://fruitgrowersnews.com/article/mating-disruption-studied-for-control-of-san-jose-scale/ (accessed on 26 October 2022).
85. Rice, R.E.; Jones, R.A. Control of San Jose scale with IGR insecticides. *ProcWOPDMC* **1999**, *73*, 89–90.
86. Sparks, T.G.; Watson, G.B.; Loso, M.R.; Geng, C.; Babcock, J.M.; Thomas, J.D. Sulfoxaflor and the sulfoximine insecticides: Chemistry, mode of action and basis for efficacy on resistant insects. *Pest. Biochem. Physiol.* **2013**, *107*, 1–7. [CrossRef]
87. Available online: https://www.canr.msu.edu/news/early-season-sprays-for-managing-san-jose-scale (accessed on 26 October 2022).
88. Bell, J.; Krueger, S.; Steffens, T. Development of Movento for sucking pest control on annual and perennial crops in the United States. *Bayer Crop. J.* **2008**, *61*, 315–328.
89. Britt, R. Evaluation of new insecticide for control of san jose scale. In Proceedings of the 82nd Annual Orchard Pest and Disease Management Conference, Hilton Hotel, Portland, OR, USA, 9–11 January 2008; Washington State University: Pullman, WA, USA, 2008; p. 59.
90. Combs, D.; Reissing, W. UltorVR 150SC for Control of New York Apple Pests. In Proceedings of the 82nd Annual Orchard Pest and Disease Management Conference, Hilton Hotel, Portland, OR, USA, 9–11 January 2007; Washington State University: Pullman, WA, USA, 2008.
91. Quesada, C.R.; Sadof, C.S. Efficacy of horticultural oil and insecticidal soap against selected armored and soft scales. *HortTechnology* **2017**, *27*, 619–624. [CrossRef]
92. Mesbah, H.A.; Abo-Shanab, A.S.; Khadiga, S.M.; Mourad, A.K.; Abdel-Razak Soad, I. Safe alternative pesticides (local mineral oils) for controlling Sjs and greedy scale insects infesting pear trees under irrigation at Burg El-Arab Area, Alexandria, Egypt. *J. Adv. Agric. Res.* **2010**, *15*, 1101–1114.
93. Masoodi, M.A.; Bhat, A.M.; Koul, V. Toxicity of insecticides to adults of Encarsia perniciosis (Hymenoptera: Aphilinidae). *Indi. J Agri. Sci.* **1988**, *59*, 50–52.
94. Fitwy, I.; Beyene., H.; Berhe, A.; Aray, A. Evaluation of some botanical extracts against major insect pests (Leafminer, Armored scale and Woolly Whitefly) of citrus plants in Central Zone of Tigray, North Ethiopia. *Momona Ethiop. J. Sci.* **2019**, *11*, 258. [CrossRef]
95. Yingfang, X.; Runqian, M.; Lloyd, S.; Steve, A. Evaluation of reduced-risk insecticides for armored scales (Hemiptera: Diaspididae) infesting ornamental plants. *J. Agric. Urban Entomol.* **2016**, *32*, 71–90. [CrossRef]
96. Salem, H.A.; Abdel-Aziz, N.F.; Sammour, E.A.; El-Bakry, A.M. Semi-field evaluation of some natural clean insecticides from essential oils on armored and soft scale insects (Homoptera: Diaspididae and Coccidae) infesting mango plants. *Int. J. ChemTech Res.* **2016**, *9*, 87–97.
97. Hansen, J.D. Ultrasound Treatments to Control Surface Pests of Fruit. *HortTechnology* **2001**, *11*, 186–188. [CrossRef]
98. Endarto, O.; Wicaksono, R.C. High-pressure water technology to control San Jose scale (*Quadraspidiotus perniciosus*) (Hemiptera: Diaspididae) on apple crops. *Adv. Biol. Sci. Res.* **2019**, *8*, 188–194. [CrossRef]
99. Angerilli, N.P.D.; Logan, D.M. The use of pheromone and barrier traps to monitor San José Scale (Homoptera: Diaspididae) phenology in the Okanagan Valley of British Columbia. *Can. Entomol.* **1986**, *118*, 767–774. [CrossRef]
100. Gaunce, A.P.; Morgan, C.V.G.; Meheriuk, M. Control of tree fruit insects with modified atmospheres. In *Controlled Atmospheres for Storage and Transport of Perishable Agricultural Commodities*; Richardson, D.G., Meheriuk, M., Eds.; Timber Press: Beaverton, OR, USA, 1981; pp. 383–390.
101. Chu, C.L. Postharvest control of San Jose Scale on apples by controlled atmosphere storage. *Postharvest Biol. Technol.* **1992**, *1*, 361–369. [CrossRef]
102. Hix, R.L. Management of San Jose scale (Homoptera: Diaspididae) on apple trees with soybean oil dormant sprays and occurrence of San Jose scale parasitoids in eastern Tennessee. Master's Thesis, University of Tennessee, Knoxville, TN, USA, 1995.

103. Argyrou, L.C. Establishment of the imported parasite *Prospaltella perniciosi* [Hym.: Aphelinidae] *Quadraspidiotus perniciosus* (Hom.: Diaspididae) in Greece. *Entqmophaga* **1981**, *26*, 125–130. [CrossRef]
104. Trandafirescu, M.; Trandafirescu, I.; Gavat, C.; Spita, V. Entomophagous complexes of some pests in apple and peach orchards in Southeastern Romania. *J. Fruit Ornam. Plant Res.* **2004**, *12*, 235–261.
105. Khan, A.A. Exploitation of Chilocorus infernalis Mulsant (Coleoptera: Coccinellidae) for suppression of the San Jose scale, *Diaspidiotus perniciosus* (Comstock) (Hemiptera: Diaspididae) in apple orchards. *J. Biolog. Contr.* **2010**, *24*, 369–372.
106. Buhroo, A.A.; Chishti, M.Z.; Masoodi, M.A. Degree-day (DD) Phenology of San Jose scale *Quadraspidiotus perniciosus* (Comstock) and the assessment of its predator, *fkhan* Mulsant in Kashmir orchard ecosystem. *Indian J. Plant Prot.* **2000**, *28*, 117–123.
107. Rawat, U.S.; Sangal, S.K.; Pawar, A.D. Biology of biocontrol of *Chilocorus bijugus* Mulsant (Coleoptera: Coccinellidae), predatory of San Jose, *Quadraspidiotus perniciosus* (Comstock). *J. Biol. Control* **1992**, *6*, 97–100.
108. Rasheed, R.; Buhroo, A.A.; Gull, S. Bionomics of *Chilocorus infernalis* Mulsant, 1853 (Coleoptera: Coccinellidae), a predator of San Jose scale, *Diaspidiotus perniciosus* (Comstock, 1881) under laboratory conditions. *Acta Agric. Slov.* **2019**, *113*, 75–81. [CrossRef]
109. Thakur, J.N.; Pawar, A.D.; Rawat, U.S. Introduction, colonisation and new records of some biological control agents of San José scale, *Quadraspidiotus perniciosus* Comstock (Hemiptera: Coccidae) in Kullu Valley, H.P., India. *J. Biol. Control.* **1993**, *7*, 99–101.
110. Thakur, J.N.; Rawat, U.S.; Pawar, A.D. Investigation on the occurance of natural enemies of San Jose scale, *Quadraspidiotus perniciosus* (Comstock) (Hemiptera: Coccidae) in J&K and Himachal Pradesh. *Entomon* **1989**, *14*, 143–146.
111. Aleksidze, G. 1995. Armored scale insects (Diaspididae), pests of fruit orchards and their control in the Republic of Georgia. *Isr. J. Entomol.* **1995**, *29*, 187–190.
112. Rawat, U.S.; Pawar, A.D. Record of natural enemies of the San Jose scale, *Quadraspidiotus perniciosus* (Comstock) from Himachal Pradesh. *J. Biol. Control* **1991**, *5*, 119–120.
113. Jolly, C.L. Biological control of San Jose scale in Himacbal Pradesh, Retrospect and Prospect. *Himachal Hortic.* **1962**, *2*, 163–168.
114. Paloukis, S. *Hispaniella lauri* (Mercet) (Hymenoptera: Aphelinidae), endoparasite of the scale insect *Quadraspidiotus perniciosus*. *Geoponika* **1979**, *246*, 29–30.
115. Hajek, A.E.; Delalibera, I., Jr. Fungal pathogens as classical biological control agents against arthropods. *BioControl* **2010**, *55*, 147–158. [CrossRef]
116. Khan, S.; Guo, L.; Maimaiti, Y.; Mijit, M.; Qiu, D. Entomopathogenic fungi as microbial biocontrol agent. *Mol. Plant Breed.* **2012**, *3*, 63–79. [CrossRef]
117. Buhroo, A.A. Biocontrol efficacy of entomopathogenic fungi against San Jose scale *Quadraspidiotus perniciosus* (Comstock) (Hemiptera: Diaspididae) in field trials. *J. Biolog. Contr.* **2014**, *28*, 214–220.

agriculture

MDPI

Article

Assessment of Biometric Parameters and Health of Canna's Cultivars as Plant Useful in Phytoremediation of Degraded Agrocenoses

Mariusz Szmagara [1], Marek Kopacki [2,*], Barbara Skwaryło-Bednarz [2], Agnieszka Jamiołkowska [2], Barbara Marcinek [1], Krystyna Rysiak [3] and Agnieszka Szmagara [4]

[1] Subdepartment of Ornamental Plants and Dendrology, Institute of Horticulture Production, University of Life Sciences in Lublin, Głęboka 28, 20-612 Lublin, Poland
[2] Department of Plant Protection, University of Life Sciences in Lublin, Leszczyńskiego 7, 20-069 Lublin, Poland
[3] Botanical Garden, Maria-Curie Sklodowska University, Sławinkowska 3, 20-810 Lublin, Poland
[4] Department of Chemistry, Institute of Biological Sciences, Faculty of Medicine, The John Paul II Catholic University of Lublin, Konstantynów 1J, 20-708 Lublin, Poland
* Correspondence: marek.kopacki@up.lublin.pl; Tel.: +48-81-524-81-18

Abstract: Recently, the ecological awareness of society and the need to take care of our surroundings and the natural environment has significantly increased. There is also an urgent problem of searching for new, environmentally friendly techniques for its purification (soil, ground and surface waters, sewage sludge and air) with the use of living organisms, especially higher plants. One plant species investigated for phytoremediation is canna. Ten varieties of canna, grown on degraded and garden soil, were tested in this respect. The disease index and species composition of fungi inhabiting its organs, growth dynamics, parameters of photosynthesis and gas exchange were determined. The conducted research showed that cannas are able to satisfactorily grow even in seemingly unfavorable soil conditions with its strong degradation. Among a total of 24 species of fungi obtained from its organs, genus *Fusarium*, considered as pathogenic for canna, *Alternaria alternata*, and, less frequently, *Thanatephorus cucumeris* and *Botrytis cinerea*, dominated. The cultivars 'Picasso', 'Cherry Red', 'President' and 'La Boheme' had lower rates of photosynthesis and gas exchange than the least affected 'Botanica', 'Wyoming', 'Robert Kemp' and 'Lucifer' cultivars. Those turned out to be the most beneficial and they can be recommended for cultivation on strongly degenerated soils.

Keywords: diseases; fungi; gas exchange; *Canna indica*; photosynthesis; transpiration

1. Introduction

In the last decade, environmental awareness has increased and counteracting environmental pollution has become the main goal of many global institutions. There is an urgent problem of searching for new, environmentally friendly techniques for purifying pollutants from water, sewage, and other contaminated materials with the use of living organisms, especially plants [1]. An interesting example is the EU-funded Phy2Climate project, which aims to validate five pilot phytoremediation in selected soil-contaminated sites, the most common in the world. The project aims to produce energy crops that will eventually power a pilot biorafinery [2]. Canna is an attractive ornamental plant often planted in urban plantings in Europe near communication routes, parking lots and other places that require covering or isolation from exhaust fumes or impurities. It is also known as a plant used in phytoremediation to purify wastewater from various pollutants that are dangerous to humans. In recent years, especially in countries with a warmer climate, canna has been increasingly used as a plant element of a sewage treatment plant useful in phytoremediation. Phosphorus and nitrogen from contaminated water from households can be effectively treated by canna [3–5], as well as lead in rhizofiltration systems [6] or oil refinery wastewater [7]. In addition, canna positively affects the preservation of biodiversity in ecosystems through flowers that attract birds and insects [8,9]. Due to its strong growth and development opportunities in coastal and flooded zones, forming large clumps in wet forests and clearings in polluted areas, it seems to be a very promising plant affecting sustainable development in rural and urban areas, despite its invasive nature in some countries [10]. In a warm climate, canna has many uses, including culinary ones, because all its parts are edible, but in the climatic conditions of Central Europe it does not survive the winter, so the rhizomes should be dug up and stored at a positive temperature [11,12]. Therefore, the possibility of using plants resistant to difficult environmental conditions, fast growing and with the ability to clean the area from pollution is sought. An important feature of these plants is a resistance to pests. In order to assess the resistance of plants to pathogens, in addition to the methods of classical mycology, the assessment of the photosynthesis process is used, which allows us to determine the condition of plants in a short time [13]. The process of photosynthesis is strictly associated with a plant species or an even variety and could be modified to a significant extend by the environmental conditions, such as temperature, precipitation and their distribution throughout the vegetation season, like numerous other parameters [14]. Concurrently, there is a major physiological plant process, which has been frequently suffered due to plant pathogens, especially those infecting the leaves [15]. The generally known fact is that leaf diseases may have destructive effect on the photosynthesis and gas exchange and can sharply reduce photosynthesis in different crop plants [16]. Pathogens cause morphological, physiological and biochemical changes, as the significant decrease in the amount of photosynthetic pigments, and thus restrict the photosynthesis rate and, resulting from it, reduce the capacity of the assimilation apparatus [17–19]. Following the plant tissues infection, many pathogens may impair photosynthesis even in the initial stage of an infection before any symptoms are visible [20,21]. The aim of the investigations was to generate an impact assessment of the infestation of the specified cultivars of canna by pathogenic fungi on the intensity of photosynthesis and the gas exchange.

2. Materials and Methods

2.1. Experimental Design

The observations were performed in 2017–2019 in the Botanical Garden in the Lublin region, Poland (51°16′ N, 22°30′ E). The most popular cultivars with a varying growth strength and high decorative qualities in the form of attractive flowers and leaves were selected for the study. These are generally available on the market and widely grown in our climatic zone. The objects of the study were ten cultivars of canna plants: Ai—'Aida', Am—'America', Bo—'Botanica', Ch—'Cherry Red', LB—'La Boheme', Lu—'Lucifer', Pi—'Picasso', Pr—'President', RK—'Robert Kemp' and Wy—'Wyoming'.

The canna rhizomes were stored for the winter period in containers covered with peat in a cool room at about 5–10 °C. The seedlings were planted into the 2 L pots in March. Then, the pots were set in a foil tunnel, in which they grew until mid-May. After the last spring frosts, they were placed in the field at 60 × 60 cm spacing.

The experiment was performed in a complete randomized blocks design with four replicates, where the block was the random effect. The experimental combination consisted of 12 plants (3 plants in 4 replicates) of each cultivar.

Mineral fertilization in accordance with the recommendations for canna plants was used.

Shoot length tests were carried out for three years from June to October. Plant height measurements were made using a measuring tape. The canna height growth figures are given as the three-year averages for all varieties.

2.2. Weather Parameters

The meteorological data sourced from the Meteorological Observatory of the Hydrology and Climatology Department, the Maria Curie-Sklodowska University in Lublin. The observations were carried out in 2017–2019 in the Botanical Garden in the Lublin region, Poland [22,23].

2.3. *Soil Analysis*

The investigations were carried out on the control (garden) and contaminated soil. The plots with contaminated soil were located near the communication route, which, apart from passers-by, was used by animal owners with their pupils. It resulted in the heavy contamination of the area with organic waste and animal excrements. So, the condition of the soil was after strong antropopressure. The control plots were with typical garden soil. The analyses of the soil in the plots were performed annually before planting the plants.

2.4. *Plant Health Assessement*

The evaluation for the level of infection disease index (DI) was performed twice every year, i.e., in mid-June and mid-September on the base of the 5-grade scale: 0—no symptoms, 1—minor spots on the leaves, 2—necrotic spots on most leaves, 3—wilting of plants and 4—dying of plants. The data were processed by McKinney's formula [24], which generates a numeric disease index (DI) of the severity of the attack: $DI = (\Sigma vn)/(NV) \times 100$, where v represents the numeric value of the class, n is the number of plants assigned to the class, N is the total number of plants in the replication and V is the numeric value of the highest class. Due to the fact that canna planted in a permanent place after the 15th May did not indicated the disease symptoms, the DI was not calculated.

2.5. *Mycological Analysis*

The presence of fungi was established at the first decade of October on the basis of etiological symptoms occurring on the infected parts of the plants and on the basis of mycological analysis being performed according to the artificial cultures method as described by Kopacki and Wagner [25]. Fungi were isolated from leaves and the stems and rhizome were analyzed in a mycological laboratory (University of Life Sciences in Lublin). Parts of the plant were pre-cleaned, and they were washed for 20 min under running water. Next, we disinfected the surface with 10% NaOCl for 60 s and then rinsed the parts three times with sterile distilled water for 3 min. Finally, the fragments were placed in a mineral medium. After separation, the obtained fungal colonies were identified to the species with the available monographs.

2.6. *Measurements of Gas Exchange Parameters*

The measurements of the photosynthetic activity of the plants were carried out in two growing seasons, in the first decade of July and first decade of September, in two combinations (plots with contaminated soil and garden soil). Ten plants per every variety have been chosen. The measurements were conducted on the 3rd fully expanded leaf counting from the base of the plant and throughout the entire growing season the same leaves were used. The following gas exchange parameters were determined: the intensity of photosynthesis (Pn) (μmol CO_2 m^{-2} s^{-1}), transpiration (E) (mmol H_2O m^{-2} s^{-1}), stomatal conductance (Gs) (mmol H_2O m^{-2} s^{-1}) and intercellular CO_2 concentration (Ci) (μmol CO_2 m^{-2} s^{-1}). They were carried out with the use of a portable infrared gas exchange analyser, CIRAS-2 PPSystems (Hitchin, Hertfordshire, UK). The analyser's cuvette conditions were set to the external source of the CO_2, humidity and temperature equal to ambient and daylight.

2.7. *Statistical Analysis*

The obtained results (for the gas exchange and disease index) were statistically analyzed with the use of an ANOVA and Tukey's confidence intervals at the 5% significance level ($\alpha = 0.05$). The Pearson correlation coefficients between photosynthesis, transpiration and the disease index was determined for both types of plots [26,27].

3. Results

3.1. Weather Parameters

During the research period, the highest temperature was recorded in the summer months, especially in July, with the exception of 2019. This year the temperature was much lower than the long-term average. The year 2017 turned out to be unusual, when a record high temperature was recorded in the autumn period. The amount of precipitation during the study period was similar to the long-term average, except for July 2019, when it was much lower than the long-term average (Figure 1).

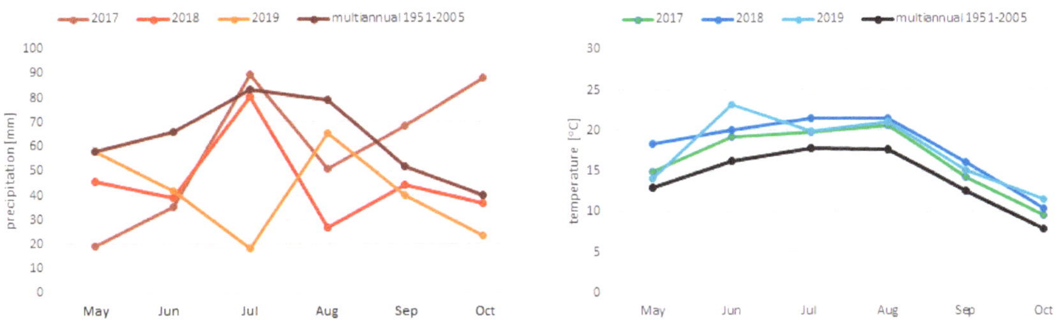

Figure 1. The average monthly rainfall and temperature.

3.2. Soil Analysis

The contaminated soil was in a bad culture, compacted by animals and had no glandular structure. It also contained numerous cat and dog droppings. The obtained results are the average of the three years of research (Table 1). The contaminated soil was characterized by a significant salinity, and the macro- and microelements contents, especially nitrogen, were significantly higher than in the garden one.

Table 1. Chemical characteristics of cultivation sites.

pH (in H_2O)	Salinity [g NaCl/L]	Macro- and Microelements Content [mg/L of Sample]									
		Contaminated Soil									
		$N\text{-}NO_3$	P	K	Ca	Mg	Zn	Mn	Cu	Fe	B
6.77	2.21	119	171	664	4524	213	23.3	5.28	4.93	39.4	2.62
		Garden Soil									
8.27	<0.24	<10.0	47	<50.0	4212	63	7.58	2.85	3.79	24	0.36

3.3. Plant Health Assessement

As a result of the research, it was observed that after planting the plants on the plots, their growth was slow. A significant increase in their length was recorded from the beginning of August. On plots with contaminated soil, a significantly better growth was recorded in three cultivars: 'Botanica', 'Robert Kemp' and 'Wyoming'. The individual plants of these varieties reached the greatest height (Figure 2). On the other hand, plants planted on plots with garden soil showed a growth dominance in the same cultivars. The growth of most varieties was very even (Figure 3).

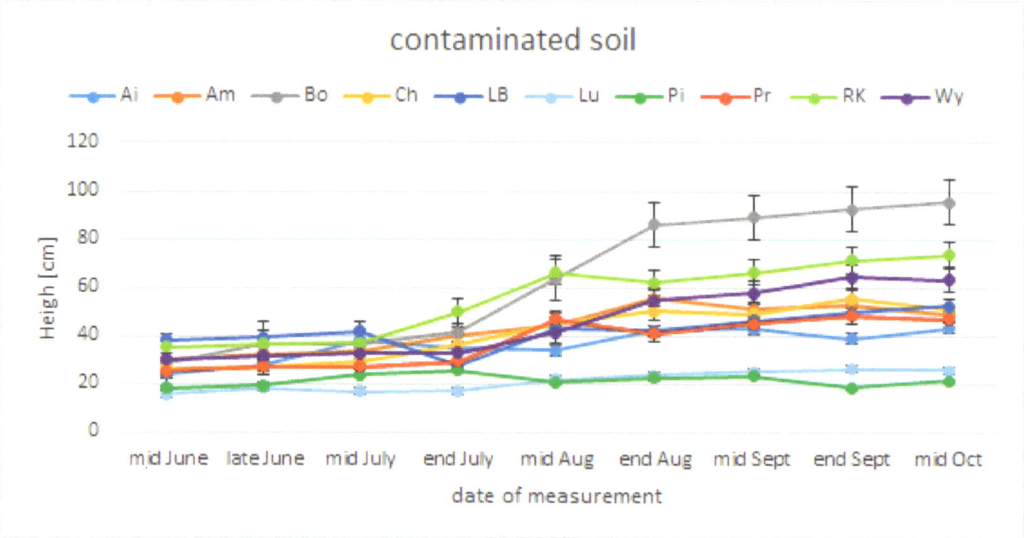

Figure 2. Dynamics of canna growth on contaminated soil. Description of cultivars is given in Materials and Methods Section 2.1. The means ± SE is expressed by each value.

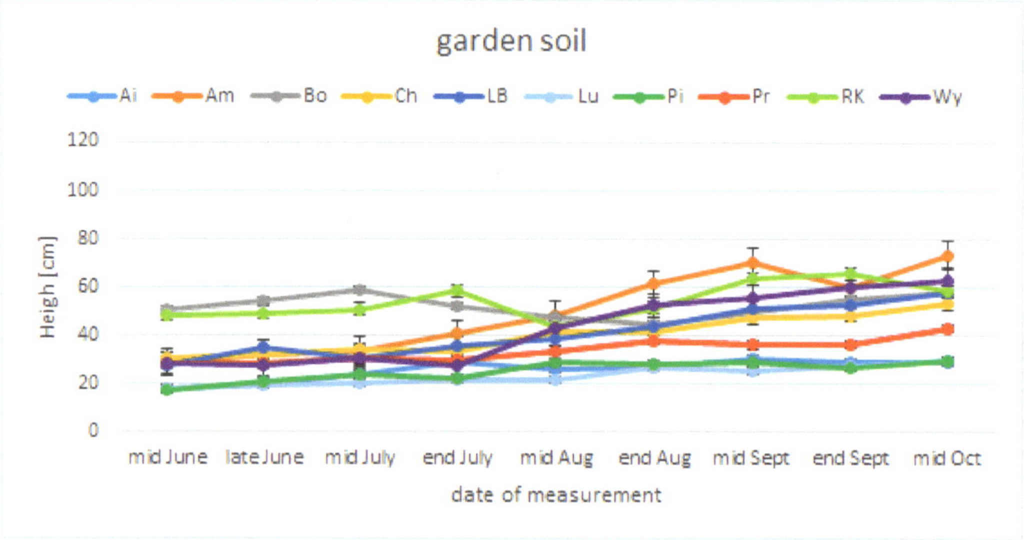

Figure 3. Dynamics of canna growth on garden soil. Description of cultivars is given in Materials and Methods Section 2.1. The means ± SE is expressed by each value.

The average disease index for all varieties was the highest in the first year of the research, but no significant statistical differences were noted. The highest index was found in plants of the variety 'Aida' (over 30%) and the lowest 'Botanica' (over 20%). In the coming years, the disease indices were significantly lower and ranged between 15 and 20% (Figure 4).

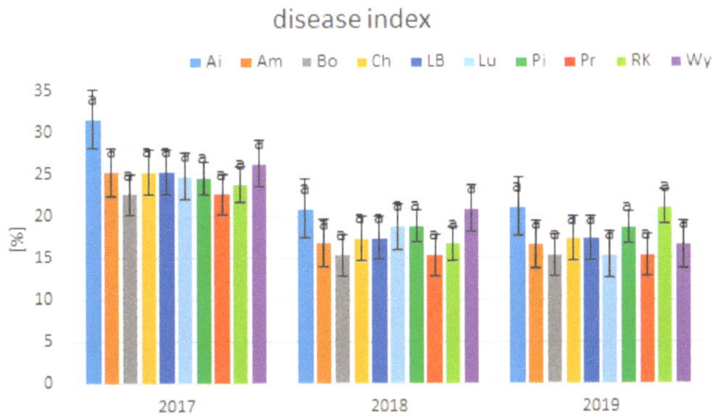

Figure 4. Disease index of canna varieties grow in 2017–2019. Description of cultivars is given in Materials and Methods Section 2.1. The means ± SE is expressed by each value. Values marked with the same letter are not significantly different (α = 0.05).

It was observed that the temperature and precipitation had an impact on fungi inhabiting canna plants. Differences between the cultivars were also observed. There were numerous yellow and brown necrotic spots visible on the leaves, which extended from the lateral veins towards the edge of the leaf blade, causing the leaves to twist over time, the leaf blade to crumble and the leaves to dry out completely. There were also, especially in the initial growth phase, extensive spots on the stems, which expanded and covered the entire stem, leading to wilting (Figure 5).

Figure 5. Disease symptoms on leaves of different canna varieties: (**A**)—'Aida', (**B**)—'America', (**C**)—'Botanica', (**D**)—'Cherry Red', (**E**)—'La Boheme', (**F**)—'Lucifer', (**G**)—'Picasso', (**H**)—'President', (**I**)—'Robert Kemp', (**J**)—'Wyoming'.

Some of the rhizomes taken out of storage completely rotted or dried up. The cross-section showed signs of conduction bundle necrosis with wet rot with brown exudate and etiological signs in the form of sporulating mycelium in grey, red, pink, yellow and brown on some rhizomes were observed. In others, however, the apical part often withered and become covered with sporulating mycelium (Figure 6).

Figure 6. Disease symptoms on rhizome of different canna varieties: (**A**)—'Aida', (**B**)—'America', (**C**)—'Botanica', (**D**)—'Cherry Red', (**E**)—'La Boheme', (**F**)—'Lucifer', (**G**)—'Picasso', (**H**)—'President', (**I**)—'Robert Kemp', (**J**)—'Wyoming'.

3.4. *Fungi Obtained from Canna Plant*

During the three-year study period (2017–2019), as a result of the mycological analysis of the canna leaves, shoots and rhizomes, a total of 5882 isolates of fungi belonging to 24 species were collected. The 2318 fungal isolates were obtained from the leaves, 2018 from the shoots and 1546 from the rhizomes (Figures 7–9).

Figure 7. Fungi colonizing leaves of canna in particular year of research. Description of cultivars is given in Materials and Methods Section 2.1.

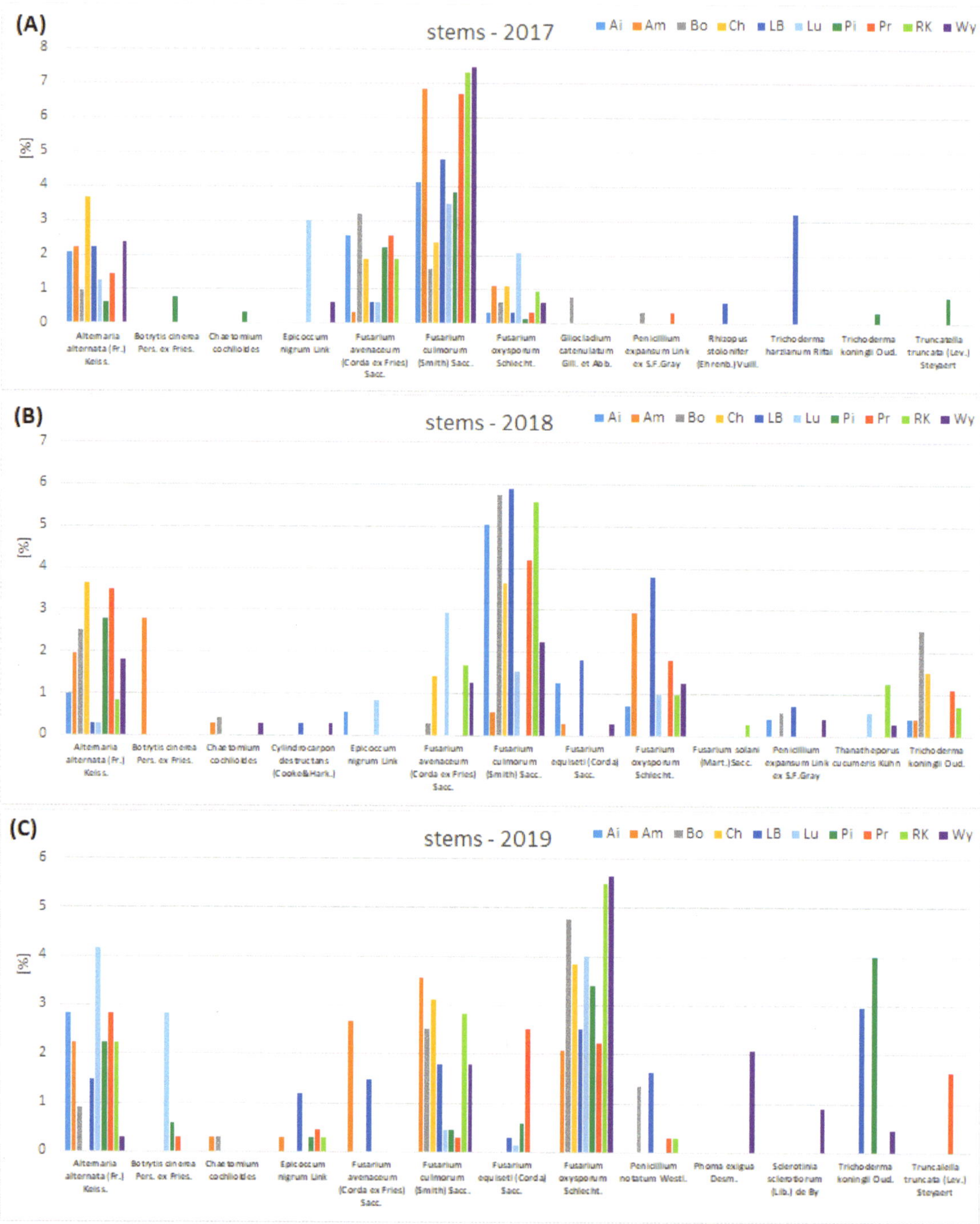

Figure 8. Fungi colonizing stems of canna in particular year of research. Description of cultivars is given in Materials and Methods Section 2.1.

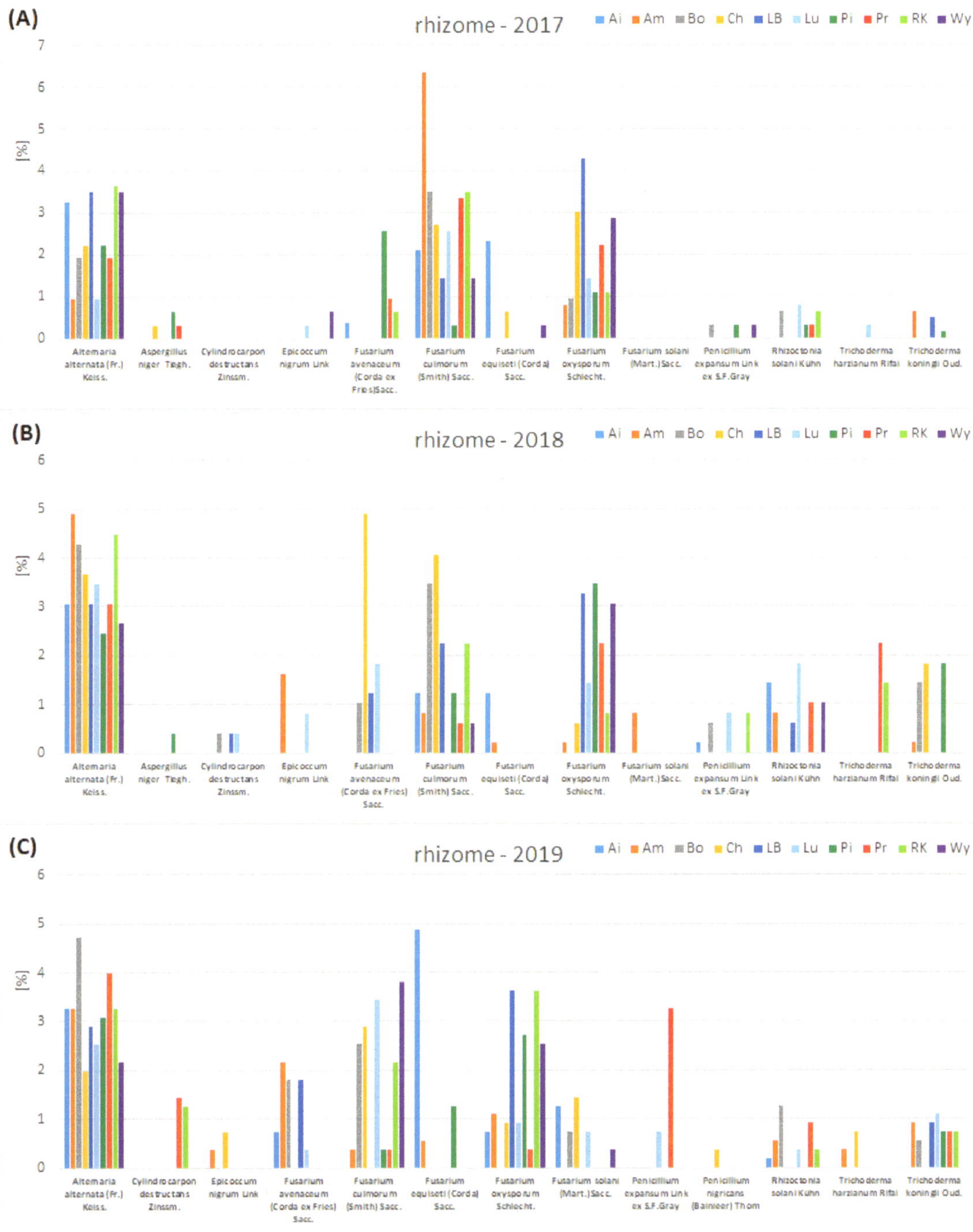

Figure 9. Fungi colonizing rhizome of canna in particular year of studies. Description of cultivars is given in Materials and Methods Section 2.1.

The pathogenic fungi of the genus *Fusarium* were the dominant colonizers of the leaves in the period of the study. They accounted for as much as 59% of the total number of fungi isolated in the first year of the study, 30% in the second year and 22% in the third year of the study (Figure 7). *F. culmorum* was most often isolated from the 'Robert Kemp' and 'America' cultivars; numerous *F. oxysporum* isolates were also obtained in the first year of research from the 'La Boheme' cultivar. During the three-year study period, the isolation of *Alternaria alternata* from all cultivars was very frequent, and accounted for 23%, 38% and 67% of the total number of fungi isolated from the leaves. They were obtained especially often from the 'President' variety in 2019. Pathogenic species *Botrytis cinerea* and *Sclerotinia sclerotiorum* were also collected from the leaves.

The most numerous canna shoots were inhabited by *A. alternata* and the fungi of the genus *Fusarium*. They were isolated from all cultivars. Most of *A. alternata* was isolated from the varieties 'Cherry Red' and 'Lucifer'. The fungi *F. culmorum*, *F. oxysporum* and *F. avenaceum* were collected in great numbers, mostly from the varieties of 'Wyoming' and 'Robert Kemp'. *B. cinerea* was often obtained from the cultivar 'America', 'Aida' and 'Picasso', and *Truncatella truncata* on 'Picasso' and 'President' were also noted (Figure 8).

Rhizomes were also colonized in great numbers by *A. alternata* and the fungi of the genus *Fusarium*, mainly *F. culmorum* and *F. avenaceum*, inhabited generally 'America' and 'Cherry Red' cultivars (Figure 3). *Cylindrocarpon obtusisporum* and *Thanatephorus cucumeris* were often isolated from the canna organs. Moreover, the occurrence of *Cylindrocarpon destructans* and *Phoma exigua* was reported (Figures 7–9).

It is noteworthy that there was the numerous colonization of all canna organs by the saprophytic fungi of the genus *Trichoderma*. From the shoots, the most often isolated was *Trichoderma koningii* in the second and the third year of study, and the isolates accounted for 10% and 7%, respectively. During the same period, numerous isolates from rhizomes were obtained and they constituted 5% and 6% of the isolated fungi. In the first and second years of the research, numerous *Epicoccum nigrum* isolates were obtained from the leaves and they accounted for 13% and 19% of the obtained fungi. Most of it was obtained from the varieties 'America' and 'La Boheme' (Figures 7–9). Useful fungus, *Chaetomium cochlioides*, was also collected from the stems during all years of studies and from the leaves in 2017 and 2018 (Figures 7 and 8).

3.5. Measurement of Gas Exchange Parameters

The analysis of the results indicated the considerable differentiation of photosynthesis and gas exchange parameters in particular canna varieties infested by pathogenic fungi. Among the determined photosynthetic intensities, the Tte lowest value, 5.61 μmol CO_2 m^{-2} s^{-1}, was noted at the 'Picasso' cultivar growing on the degraded substrate and the obtained results differed significantly from other cultivars (Table 2). Similar low results were reported at cultivars 'Cherry Red' and 'President', 8.90 and 9.78 μmol CO_2 m^{-2} s^{-1}, respectively. The analogous trend was observed also on the uncontaminated plots. The lowest level of photosynthesis was carried out at 'Picasso', 6.95, and 'La Boheme', 7.17 CO_2 m^{-2} s^{-1}, and the obtained results differ significantly from other cultivars, except for 'America' and 'Robert Kemp', at which values 7.77 and 8.63 μmol CO_2 m^{-2} s^{-1} were noted. Nevertheless, the highest photosynthesis intensities were observed in 'Botanica', 'Wyoming' and 'Lucifer' varieties (Table 2).

The studied varieties have also differed in their transpiration rate (E). On contaminated soil, the lowest transpiration was reported in the 'Aida' cultivar, 1.76 mmol H_2O m^{-2} s^{-1}. The other cultivars carried the transpiration out at the level of ca. 2 mmol H_2O m^{-2} s^{-1}, except for 'America', which evaporated the most water, 3.18 mmol H_2O m^{-2} s^{-1}, from the leaf area, and this value differs significantly from all other cultivars. The similar tendency continued also in the plots with garden soil (Table 2).

Table 2. The photosynthesis intensity and gas exchange parameters of the canna at various cultivation sites.

Variety	Pn	E	Gs	Ci
	Contaminated Soil			
Aida	10.56 b–f	1.76 g	94.50 f	304.67 efg
America	11.06 a–e	3.18 a	152.50 ab	392.25 b–e
Botanica	14.12 a	2.01 fg	119.67 b–f	502.50 ab
Cherry Red	8.90 d–g	2.38 c–g	90.00 f	393.67 a–e
La Boheme	11.31 a–e	2.20 d–g	106.25 def	333.75 d–g
Lucifer	12.23 abc	2.18 d–g	113.50 c–f	456.08 abc
Picasso	5.61 h	2.09 efg	103.67 def	364.25 c–f
President	9.78 c–g	2.83 c–g	104.25 def	412.25 a–e
Robert Kemp	12.32 abc	2.42 b–f	123.25 b–f	508.08 a
Wyoming	12.08 a–d	2.07 efg	96.08 ef	337.67 d–g
Mean	10.80 a	2.31 b	110.37 b	400.52 a
	Garden Soil			
Aida	9.43 c–g	2.52 b–f	145.33 abc	305.67 efg
America	7.77 fgh	3.05 ab	151.17 ab	360.67 c–f
Botanica	13.23 abc	2.67 a–e	164.83 a	440.17 a–d
Cherry Red	10.63 b–f	3.03 abc	142.83 abc	360.00 c–f
La Boheme	7.17 gh	2.63 a–f	129.50 b–e	255.17 fg
Lucifer	11.18 a–e	2.33 d–g	112.33 c–f	355.83 c–f
Picasso	6.95 gh	2.57 a–f	122.17 b–f	337.00 d–g
President	9.35 c–g	2.67 a–e	133.33 a–d	342.00 c–g
Robert Kemp	8.63 e–h	2.83 a–d	146.33 abc	442.00 c–g
Wyoming	10.58 b–f	2.26 d–g	98.17 ef	240.83 g
Mean	9.49 b	2.65 a	134.60 a	343.98 b

Pn—photosynthesis (μmol CO_2 m^{-2} s^{-1}), E—transpiration rate (mmol H_2O m^{-2} s^{-1}), Gs—stomatal conductance (mmol H_2O m^{-2} s^{-1}), Ci—intercellular carbon dioxide concentration (μmol CO_2 m^{-2} s^{-1}). Values designated with the same letter are not significantly different ($\alpha = 0.05$).

The lowest stomatal conductance (Gs) in relation to the other varieties was shown by 'Cherry Red', 90.00 mmol H_2O m^{-2} s^{-1}, on plots with contaminated soil and 'Aida' by 94.50 mmol H_2O m^{-2} s^{-1}. In turn, 'Wyoming', during the entire period of the measurements, was in the range of 96.08–98.17 mmol H_2O m^{-2} s^{-1}. The 'Picasso', 'La Boheme' and 'Lucifer' varieties also showed a lower level of stomatal conductance in both sites. The highest transpiration in all combinations was showed by the varieties 'Botanica', 'America' and 'Robert Kemp' (Table 2).

The lowest concentration of intercellular carbon dioxide (Ci) compared to other varieties during the observations in all combinations was recorded in the cultivars 'Aida', 'La Boheme', 'Wyoming' and 'Picasso' and ranged from 240.83 to 364.25 μmol CO_2 m^{-2} s^{-1}. Nevertheless, the highest Ci values have been notified in the cultivars 'Robert Kemp' and 'Botanica' and ranged between 440.17 and 508.08 μmol CO_2 m^{-2} s^{-1} (Table 2).

4. Discussion

The obtained results are consistent with those of other authors. Different species and varieties of plants react to a different degree of infection by pests, which manifests itself in interfering with the physiological processes in the plant. In the majority of diseases, the photosynthesis rate may be reduced at the beginning of the infection. Indeed, most pathogens decreased their photosynthesis levels from the onset of infection even though there were no visible symptoms [19,20,28].

The low photosynthesis parameters are correlated with a high air temperature during the growing season and a limited and unevenly distributed rainfall. The photosynthesis process is uniquely connected with unfavorable meteorological conditions due to the acute sensitivity of the photosynthetic apparatus in the assimilation organs. In the response to a water deficit, the stomata closes, which then results in a decrease in the stomatal conductance, thus the limiting the availability of CO_2, consequently resulting in the reduction in the photosynthetic intensity [14]. In the conducted research, this was particularly evident in 2017, when the weather parameters differed from the long-term averages.

The examined varieties differed also in the transpiration rate (E). In the first year of studies, the lowest transpiration was reported in the 'Aida' cultivar. The other cultivars carried the transpiration out at the similar level, except for 'America', which evaporated the most water, 3.18 mmol H_2O m^{-2} s^{-1} from the leaf area, and this value differed significantly from all other cultivars. A similar tendency continued also in the second combination. In the studies of Lobato et al. [17], it was confirmed that the transpiration rate corresponded to the plant infection degree and is lower in the infested plants.

The lowest stomatal conductance (Gs) among all cultivars on the contamination fields was showed at 'Cherry Red' and 'Aida'. However, the cultivar 'Wyoming' has stomatal conductance (Gs) in the range of 96.08–98.17 mmol H_2O m^{-2} s^{-1}. The stomatal conductance at a low level have also 'Picasso', 'La Boheme' and 'Lucifer' cultivars. The highest values of transpiration during the performed studies were noticed in 'Botanica', 'America' and 'Robert Kemp' cultivars. In conclusion, higher photosynthesis parameters were found in plants growing on contaminated soil. Numerous researchers have observed a similar trend on other plant species. The investigations of Bispo et al. [28] have confirmed that *Ceratocystis fimbriata* isolates caused the decrease in stomatal conductance independently of the mango variety, and therefore simultaneously they cause a diffusive limitation of the assimilation of CO_2. It was proved that the reduced level of Gs is the one of major limitations of photosynthesis in infected plants by lowering the availability of the CO_2 flow by the leaf area [19]. The research of Ribeiro et al. [29] is concur with ours and shows that the factors of the CO_2 assimilation and stomatal conductance (Gs) were higher on the healthy seedlings of orange than on those infested by *Xylella fastidiosa*. Polanco et al. [16] has also demonstrated an analogous relationship in bean plants infested by *Colletotrichum lindemuthianum*.

The lowest intercellular CO_2 (Ci) values, in comparison to other cultivars, were noticed in 'Aida', 'La Boheme', 'Wyoming' and 'Picasso'. However, the highest values were found in 'Robert Kemp' and 'Botanica' cultivars by Mikiciuk et al. [30]. They had demonstrated that Ci could indicate a significant variation depending on the physiological and developmental stage of a plant, and is higher at the beginning of the growing season than at the end. This tendency is also proved in our research.

Analyzing the level of DI, no statistically significant differences were noted between the levels, but the level of infection was quite low. It is possible that at a higher level these differences would be more noticeable. Observing the colonization of plants by fungi in a polluted field, *Alternaria alternata* and the fungi of the genus *Fusarium* dominated. The number of populations of these fungi varied in particular years of research and could be related to the weather conditions, especially the soil moisture, which affects the development of these fungi [31,32]. The polyphagous *Fusarium oxysporum*, which often inhabits canna, is grown in warm climates [33,34]. Isolated *Fusarium culmorum* and *Thanatephorus cucumeris* are known as the cause of shoot base and rhizome rot, which often occur on numerous species of ornamental plants [35]. High humidity is associated with the occurrence of *S. sclerotiorum*, causing the rotting of many plant species [36]. An often isolated in tropical countries, weak pathogen, *Alternaria alternata*, has in recent years been frequently noted on canna plants used for wastewater treatment [37]. This fungus is responsible for considerable losses in the production of canna in tropical countries [38]. Numerous populations of *A. alternata* obtained in our investigation suggest further studies on its pathogenicity to canna. Frequently isolated antagonistic fungi, like *Trichoderma* sp. or *Chaetomium* sp. and *Epicoccum nigrum*, have a great influence on the health of canna plants due to the reduction in the pathogenic fungi number, especially in the soil [39–42]. Currently, the contamination of the environment, especially areas after anthropopressure with heavy metals and other pollutions, is a serious problem. An effective way to purify it is phytoremediation. Research conducted by Trąmpczyńska and Gawroński [43] in Poland proved that the planting of garden canna in the area of urban greenery contributed to the reduction in the lead contamination of the planted stand. Many authors confirm that canna can be a very valuable plant to purify polluted water in different climatic zones ([1,44–46]) and remove pesticides from the environment [47]. Taking into account the climate change and the assumptions of the "European Green Deal" recently introduced, it can be assumed that canna will become a plant recommended for planting biological sewage treatment plants.

5. Conclusions

Canna plants were most often inhabited by the fungi of the genus *Fusarium*, considered as being pathogenic to canna and *Alternaria alternata*, and less frequently by *Thanatephorus cucumeris* and *Botrytis cinerea*.

'Picasso', 'Cherry Red', 'President' and 'La Boheme', as the more infected by pathogens cultivars, carried out the photosynthesis and gas exchange processes on significantly lower levels than the less infested cultivars of 'Botanica', 'Wyoming', 'Robert Kemp' and 'Lucifer'.

In the conducted research, the cultivation of the canna cultivars 'Botanica', 'Robert Kemp' and 'Wyoming' in difficult conditions on contaminated soils turned out to be the most beneficial, so they can be recommended for cultivation on strongly degenerated soils.

Author Contributions: Conceptualization, M.S. and M.K.; methodology, M.K., M.S. and B.S.-B.; software, M.S., A.J. and B.M.; validation, M.S., M.K. and B.S.-B.; formal analysis, M.S. and M.K. and A.J.; investigation, M.K., M.S., B.M. and K.R.; resources, K.R.; data curation, M.S. and M.K.; writing—original draft preparation, M.S. and M.K.; writing—review and editing, M.S., M.K. and A.S.; visualization, M.K. and A.S.; supervision and project administration, M.S. and M.K. All authors have read and agreed to the published version of the manuscript.

Funding: This research received no external funding.

Institutional Review Board Statement: Not applicable.

Data Availability Statement: Data are available upon reasonable request to the corresponding author.

Conflicts of Interest: The authors declare no conflict of interest.

References

1. Hassan, I.; Chowdhury, S.R.; Prihartato, P.K.; Razzak, S.A. Wastewater Treatment Using Constructed Wetland: Current Trends and Future Potential. *Processes* **2021**, *9*, 1917. [CrossRef]
2. European Commission. *A Global Approach for Recovery of Arable Land through Improved Phytoremediation Coupled with Advanced Liquid Biofuel Production and Climate Friendly Copper Smelting Process*; Horizon 2020; European Commission: Luxembourg, 2020.
3. Ojoawo, S.O.; Udayakumar, G.; Naik, P. Phytoremediation of Phosphorus and Nitrogen with Canna x Generalis Reeds in Domestic Wastewater through NMAMIT Constructed Wetland. *Aquat. Procedia* **2015**, *4*, 349–356. [CrossRef]
4. Arliyani, I.; Tangahu, B.; Mangkoedihardjo, S. Plant Diversity in a Constructed Wetland for Pollutant Parameter Processing on Leachate: A Review. *J. Ecol. Eng.* **2021**, *22*, 240–255. [CrossRef]
5. Plaza, B.M.; Maggini, R.; Borghesi, E.; Pardossi, A.; Lao, M.T.; Jiménez-Becker, S. Nutrient Extraction in Pansy Fertigated with Pure, Diluted, Depurated and Phytodepurated Leachates from Municipal Solid Waste. *Agronomy* **2020**, *10*, 1911. [CrossRef]
6. Čule, N.; Vilotic, D.; Nešić, M.; Veselinovic, M.; Drazic, D.; Mitrovic, S. Phytoremediation Potential of Canna Indica L. in Water Contaminated with Lead. *Fresenesius Environ. Bull.* **2016**, *25*, 3728–3733.
7. Ghezali, K.; Bentahar, N.; Barsan, N.; Nedeff, V.; Moșneguțu, E. Potential of Canna Indica in Vertical Flow Constructed Wetlands for Heavy Metals and Nitrogen Removal from Algiers Refinery Wastewater. *Sustainability* **2022**, *14*, 4394. [CrossRef]
8. Glinos, E.; Cocucci, A.A. Pollination Biology of Canna Indica (Cannaceae) with Particular Reference to the Functional Morphology of the Style. *Plant Syst. Evol.* **2011**, *291*, 49–58. [CrossRef]
9. Yeo, P. *Secondary Pollen Presentation: Form, Function, and Evolution*; Plant Systematics and Evolution; Springer: Wien, Austria; New York, NY, USA, 1993; ISBN 978-0-387-82448-2.
10. Henderson, L.; Cilliers, C.J. *Invasive Aquatic Plants—A Guide to the Identification of the Most Important and Potentially Dangerous Invasive Aquatic and Wetland Plants in South Africa*; ARC-Plant Protection Research Institute: Pretoria, South Africa, 2002.
11. Okonwu, K.; Ariaga, C. Nutritional Evaluation of Various Parts of *Canna Indica* L. *ARRB* **2016**, *11*, 1–5. [CrossRef]
12. Piperno, D.R. The Origins of Plant Cultivation and Domestication in the New World Tropics: Patterns, Process, and New Developments. *Curr. Anthropol.* **2011**, *52*, S453–S470. [CrossRef]

13. Wagner, A.; Jamiołkowska, A.; Michałek, W. Pathogenicity of *Fusarium Oxysporum* from Different Soil Environments and Its Effect on Photosynthetic Activity of Tomato Plants. *EJPAU* **2007**, *10*, 29.
14. Olszewska, M.; Grzegorczyk, S.; Olszewski, J.; Bałuch-Małecka, A. A Comparison of the Response of Selected Grass Species to Water Stress. *Grassl. Sci. Pol.* **2010**, *13*, 127–137.
15. Rios, V.S.; Rios, J.A.; Aucique-Pérez, C.E.; Silveira, P.R.; Barros, A.V.; Rodrigues, F.Á. Leaf Gas Exchange and Chlorophyll *a* Fluorescence in Soybean Leaves Infected by *Phakopsora Pachyrhizi*. *J. Phytopathol.* **2018**, *166*, 75–85. [CrossRef]
16. Polanco, L.R.; Rodrigues, F.A.; Nascimento, K.J.T.; Cruz, M.F.A.; Curvelo, C.R.S.; DaMatta, F.M.; Vale, F.X.R. Photosynthetic Gas Exchange and Antioxidative System in Common Bean Plants Infected by Colletotrichum Lindemuthianum and Supplied with Silicon. *Trop. Plant Pathol.* **2014**, *39*, 35–42. [CrossRef]
17. Lobato, A.; Gonçalves-Vidigal, M.; Vidigal Filho, P.; Andrade, C.; Kvitschal, M.; Bonato, C. Relationships between Leaf Pigments and Photosynthesis in Common Bean Plants Infected by Anthracnose. *N. Z. J. Crop Hortic. Sci.* **2010**, *38*, 29–37. [CrossRef]
18. Lobato, A.K.S.; Mc, G.-V.; PS Vidigal, F.; Costa, R.C.L.; Cruz, F.J.R.; Santos, D.G.C.; Silva, C.R.; Li, S.; Ll, S. Changes in Photosynthetic Pigment and Carbohydrate Content in Common Bean Cultivars Infected by *Colletotrichum Lindemuthianum*. *Plant Soil Environ.* **2009**, *55*, 58–61. [CrossRef]
19. Alves, A.A.; Guimarães, L.M.d.S.; Chaves, A.R.d.M.; DaMatta, F.M.; Alfenas, A.C. Leaf Gas Exchange and Chlorophyll a Fluorescence of Eucalyptus Urophylla in Response to Puccinia Psidii Infection. *Acta Physiol. Plant* **2011**, *33*, 1831–1839. [CrossRef]
20. Berger, S.; Sinha, A.K.; Roitsch, T. Plant Physiology Meets Phytopathology: Plant Primary Metabolism and Plant Pathogen Interactions. *J. Exp. Bot.* **2007**, *58*, 4019–4026. [CrossRef] [PubMed]
21. Moshou, D.; Bravo, C.; Oberti, R.; West, J.; Bodria, L.; McCartney, A.; Ramon, H. Plant Disease Detection Based on Data Fusion of Hyper-Spectral and Multi-Spectral Fluorescence Imaging Using Kohonen Maps. *Real-Time Imaging* **2005**, *11*, 75–83. [CrossRef]
22. Hortus Botanicus Universitatis Mariae Curie-Skłodowska. *Index Seminum 2018*; Hortus Botanicus Universitatis Mariae Curie-Skłodowska: Lublin, Polonia, 2019.
23. Hortus Botanicus Universitatis Mariae Curie-Skłodowska. *Index Seminum 2019*; Hortus Botanicus Universitatis Mariae Curie-Skłodowska: Lublin, Polonia, 2020.
24. Parafiniuk, S.; Kopacki, M. Biological Efficacy of the Chemical Chrysanthemums Protection with the Use of Fine and Coarse Droplets. *J. Cent. Eur. Agric.* **2012**, *13*, 554–559. [CrossRef]
25. Kopacki, M.; Wagner, A. Effect of Some Fungicides on Mycelium Growth of Fusarium Avenaceum (Fr.) Sacc. Pathogenic to Chrysanthemum (Dendranthema Grandiflora Tzvelev). *Agron. Res.* **2006**, *4*, 237–240.
26. Bordens, K.S.; Abbott, B.B. *Research Design and Methods. A Process Approach*, 7th ed.; McGraw-Hill: New York, NY, USA, 2008.
27. Raudonius, S. Application of Statistics in Plant and Crop Research: Important Issues. *Zemdirb. Agric.* **2017**, *104*, 377–382. [CrossRef]
28. Bispo, W.M.d.S.; Araujo, L.; Moreira, W.R.; Silva, L.d.C.; Rodrigues, F.Á. Differential Leaf Gas Exchange Performance of Mango Cultivars Infected by Different Isolates of Ceratocystis Fimbriata. *Sci. Agric. (Piracicaba Braz.)* **2016**, *73*, 150–158. [CrossRef]
29. Ribeiro, R.V.; Machado, E.C.; Oliveira, R.F. Growth- and Leaf-Temperature Effects on Photosynthesis of Sweet Orange Seedlings Infected with Xylella Fastidiosa. *Plant Pathol.* **2004**, *53*, 334–340. [CrossRef]
30. Mikiciuk, G.; Mikiciuk, M.; Ptak, P. The Effect of Antitranspirant Di-1-p-Methene on Some Physiological Traits of Strawberry. *J. Ecol. Eng.* **2015**, *16*, 161–167. [CrossRef]
31. Kopacki, M.; Wagner, A. Pathogenicity of Fusarium Spp. to Chrysanthemum (*Dendranthema Grandiflora* Tzvelev). *Latv. J. Agron.* **2004**, *7*, 158–159.
32. Jerushalmi, S.; Maymon, M.; Dombrovsky, A.; Freeman, S. Fungal Pathogens Affecting the Production and Quality of Medical Cannabis in Israel. *Plants* **2020**, *9*, 882. [CrossRef]
33. Kunal, T.; Milind, G.; Shrirang, B. Diversity of Rhizosphere Mycoflora of *Canna Indica* L. *Int. J. Curr. Res. Life Sci.* **2018**, *7*, 1669–1671.
34. Olivares, B.O.; Rey, J.C.; Lobo, D.; Navas-Cortés, J.A.; Gómez, J.A.; Landa, B.B. Fusarium Wilt of Bananas: A Review of Agro-Environmental Factors in the Venezuelan Production System Affecting Its Development. *Agronomy* **2021**, *11*, 986. [CrossRef]
35. Sharma, S.; Chandel, S. Management of Stem Rot (*Rhizoctonia Solani*) of Carnation by Fungicides. *J. Mycol. Pl. Pathol.* **2013**, *43*, 187–189.
36. Grabowski, M.A.; Malvick, D.K. Evaluation of Ornamental Tropical Plants for Resistance to White Mold Caused by *Sclerotinia Sclerotiorum*. *Hortscience* **2017**, *52*, 1375–1379. [CrossRef]
37. Kundu, D.; John, D.J.; Adhikari, T.; Ghosh, P.; Sarkar, S.; Mitra, A.K. Study of Rhizospheric Association in Improving the Effectiveness of a Phytorid Plant towards Bioremediation. *World J. Pharm. Res.* **2016**, *5*, 1546–1556.
38. Roopa, P.; Fugro, P.A.; Kadam, J. Symptomatology, Host Range Study and Management by Botanicals against *Alternaria Alternata* of *Canna Indica* (Fr.) Keissler. *Int. J. Life Sc. Bt. Pharm. Res.* **2014**, *3*, 116–120.
39. Ogórek, R.; Plaskowska, E. Epicoccum Nigrum for Biocontrol Agents in Vitro of Plant Fungal Pathogens. *Commun. Agric. Appl. Biol. Sci.* **2011**, *76*, 691–697.
40. Fatima, N.; Ismail, T.; Muhammad, S.A.; Jadoon, M.; Ahmed, S.; Azhar, S.; Mumtaz, A. *Epicoccum* Sp. an Emerging Source of Unique Bioactive Metabolites. *Acta Pol. Pharm.* **2016**, *73*, 13–21.
41. Soytong, K.; Kahonokmedhakul, S.; Song, J.; Tongon, R. Chaetomium Application in Agriculture. In *Technology in Agriculture*; Ahmad, F., Sultan, M., Eds.; IntechOpen: London, UK, 2021; ISBN 978-1-83881-921-7.

42. Tyśkiewicz, R.; Nowak, A.; Ozimek, E.; Jaroszuk-Ściseł, J. Trichoderma: The Current Status of Its Application in Agriculture for the Biocontrol of Fungal Phytopathogens and Stimulation of Plant Growth. *Int. J. Mol. Sci.* **2022**, *23*, 2329. [CrossRef]
43. Trąmpczyńska, A.; Gawroński, S.W. Canna x Generalis Jako Fitoremediant Terenów Zurbanizowanych. *Folia Hortic.* **2003**, *1*, 447–449.
44. Choudhary, A.; Kumar, S.; Sharma, C.; Kumar, P. Performance of Constructed Wetland for the Treatment of Pulp and Paper Mill Wastewater. In Proceedings of the World Environmental and Water Resources Congress, Palm Springs, CA, USA, 22–26 May 2011.
45. Huang, T.; Liu, W.; Zhang, Y.; Zhou, Q.; Wu, Z.; He, F. A Stable Simultaneous Anammox, Denitrifying Anaerobic Methane Oxidation and Denitrification Process in Integrated Vertical Constructed Wetlands for Slightly Polluted Wastewater. *Environ. Pollut.* **2020**, *262*, 114363. [CrossRef] [PubMed]
46. Karungamye, P.N. Potential of Canna Indica in Constructed Wetlands for Wastewater Treatment: A Review. *Conservation* **2022**, *2*, 499–513. [CrossRef]
47. Chen, Q.; Zeng, H.; Liang, Y.; Qin, L.; Peng, G.; Huang, L.; Song, X. Purification Effects on β-HCH Removal and Bacterial Community Differences of Vertical-Flow Constructed Wetlands with Different Vegetation Plantations. *Sustainability* **2021**, *13*, 13244. [CrossRef]

agriculture

MDPI

Article

Forecasting of Hypoallergenic Wheat Productivity Based on Unmanned Aerial Vehicles Remote Sensing Approach—Case Study

Bogdan Kulig [1], Jacek Waga [2], Andrzej Oleksy [1], Marcin Rapacz [2], Marek Kołodziejczyk [1], Piotr Wężyk [3], Agnieszka Klimek-Kopyra [1,*], Robert Witkowicz [1], Andrzej Skoczowski [4], Grażyna Podolska [5] and Wiesław Grygierzec [6]

1 Department of Agroecology and Crop Production, University of Agriculture in Krakow, 31-120 Krakow, Poland
2 Department of Physiology, Plant Breeding and Seed Production, University of Agriculture in Krakow, 30-239 Krakow, Poland
3 Department of Forest Resource Management, Faculty of Forestry, University of Agriculture in Krakow, 31-425 Krakow, Poland
4 The Franciszek Górski Institute of Plant Physiology, Polish Academy of Sciences, 31-342 Krakow, Poland
5 Cereal Crop Department, Institute of Soil Science and Plant Cultivation, 24-100 Puławy, Poland
6 Department of Statistics and Social Policy, University of Agriculture in Krakow, 31-120 Krakow, Poland
* Correspondence: agnieszka.klimek@urk.edu.pl

Abstract: Remote sensing methods based on UAV and hand-held devices as well have been used to assess the response to nitrogen and sulfur fertilization of hypoallergenic genotypes of winter wheat. The field experiment was conducted using the split-split-plot design with three repetitions. The first factor was the two genotypes of winter wheat specified as V1 (without allergic protein) and V2 (with allergic protein), and the second factor was three doses of sulfur fertilization: 0, 20 and 40 kg S per ha. The third factor consisted of six doses of nitrogen fertilization: 0, 40, 60, 80, 100 and 120 kg N ha^{-1}. Monitoring the values of the indicators depending on the level of nitrogen and sulfur fertilization allowed the results to be used in yield forecasting, assessment of plant condition, LAI value, nutritional status in the cultivation of wheat. The maximum yield should be expected at doses of 94 and 101 kg N ha^{-1} for genotypes V1 and V2, respectively, giving yields of 5.39 and 4.71 Mg ha^{-1}. On the basis of the tested vegetation indices, the highest doses of N should be applied using the normalized difference RedEdge (NDRE), and the lowest ones based on the enhanced vegetation index (EVI), and, in the latter case, a reduction in yield of more than 200 kg ha^{-1} in the V2 genotype should be taken into account.

Keywords: hypoallergenic wheat; yield forecasting; vegetation indices; GHG; UAV

1. Introduction

Wheat is one of the most important crops used by humans for food production. In terms of the size of the harvested grain yield, it ranks second in the world, directly after maize. Wheat flour is the only raw material whose protein components have the potential to create gluten—a protein substance that determines its baking properties. Gluten is made up of two highly polymorphic groups of proteins—gliadins and glutenins. The differentiation of the physicochemical structure of their fractions and subunits is genetically determined [1]. Despite the significant differences, gliadin and glutenin also have a number of common traits, including the most prominent share of proline and glutamine, which account for 17 and 40% of all amino acids, respectively. They are therefore referred to as prolamins [2]. Both groups of proteins make up more than 80% of gluten by weight [3]. However, in addition to their beneficial functional properties, as food allergens, prolamin proteins may negatively affect human health. In the case of people allergic to gluten, they cause a number of diseases, such as gluten enteropathy, atopic dermatitis (Dühring's disease), urticaria, asthma, angioedema, food allergy to gluten, gluten ataxia, non-celiac gluten sensitivity and even anaphylactic shock directly threatening life (wheat-dependent exercise-induced anaphylaxis—WDEIA) [4–7].

The work of Waga and Skoczowski [8] resulted in the creation of winter wheat hybrid lines V1 (wasko.gl−) with a significantly reduced number of highly allergenic gliadin proteins from the ω group and low molecular weight glutenins (the so-called D-type LMW glutenin). These lines were created by traditional breeding methods based on combinations of crosses and selection supported by electrophoretic studies of gliadin and glutenin proteins. Basic agronomical research [9] has already been conducted to verify the impact of N-fertilization on the protein structure of hypoallergenic wheat genotypes. However, no essential field studies have been carried out to assess the effect of N-fertilization of new hypoallergenic wheat genotypes on the vegetation indices values measured by remote sensing techniques.

Remote sensing research consists in acquiring, processing and interpreting data characterizing the tested object in terms of the amount of reflected or emitted electromagnetic radiation [10–12]. Different spectral responses by plants can be acquired by sensors (e.g., multispectral, thermal or hyperspectral) mounted on airborne (aircraft; UAV—unmanned aerial vehicles), satellite (e.g., SENTINEL-2 ESA) or ground-based devices (e.g., field spectrometers).

Remote sensing of crops based on different scale-level remote sensing sensors (from UAV, trough aerial and satellite) has been improved by applied research and state-of-the art technology over the last years [10–12]. The innovative unmanned aerial vehicles (UAVs) seems to be the most advanced tool for high resolution (spatial, spectral, radiometrical and time) data collection as a monitoring platform for smart precision agriculture, and sometimes as a tool for direct treatments (e.g., spreading herbicides) [10–12].

Using UAVs, digital imagery can achieve a spatial resolution (GSD—ground sample distance) of a single cm (e.g., enabling researchers to count the plant density or detect diseases) helping precision agriculture applications in the monitoring of vegetation growth. The UAV platforms (also called UAS) can be equipped with different digital cameras, such as: high-resolution RGB (e.g., 45 Mpx), multispectral (e.g., 5 bands), hyperspectral (e.g., 220 bands) or thermal. In addition to spectral information, the 3D point cloud based on LiDAR sensors can be acquired, delivering much crucial information about the biometric data (e.g., height or cover density of the canopy). Such sensors allow the monitoring of changes happening during the vegetation season due to plant growth phases and different limitation factors (diseases, drought, wind etc.). In many studies the ground-truth and UAV collected information was useful for the better tuning of spectral reflection acquired by satellite remote sensing sensors, covering wide areas [10–12]. Such an approach based on spectral information enables the development of many so-called vegetation indices (e.g., LAI, NDVI) used for mathematical models describing, e.g., soil moisture, health conditions or approximated yield. According to Huete and Justice [13], the vegetation index should be highly correlated with the biophysical parameters of plants, the most common of which are: biomass, leaf area index (LAI) and absorbed photosynthetically active radiation (PAR). The most commonly used vegetation index is the NDVI (normalized difference vegetation index). In measurements of the vegetation state, the NDVI values most often range from 0.0 to +0.8 [14–17].

Scientific research has shown a strong correlation between the LAI value and the vegetation indices, with differentiation in the strength of the correlation with respect to different indices. In turn, the LAI index depends on the phase of plant development and the applied agrotechnical treatments, in particular nitrogen and sulfur fertilization, which affect the efficiency of nitrogen application [18]. Many model experiments on the fertilization of wheat cultivars have been carried out [19–24]. However, due to the need to search for genotypes that are less allergenic to people, there is a return to the older type of varieties, characterized by lower requirements as to the level of nitrogen fertilization. Monitoring the values of the indicators depending on the level of nitrogen and sulfur fertilization allows the results to be used in yield forecasting, assessment of plant condition, LAI value, nutritional status and stress conditions in the cultivation of hypoallergenic wheat. The null hypothesis assumes that a genotype containing the complete set of allergenic proteins and one which is completely devoid of these proteins will show similar values of vegetation indices and a similar reaction to a specific technological path. The purpose of this study was to determine the usefulness of selected vegetation indices obtained using a remote method to assess the yield, nutritional status and fertilization needs of the two specific genotypes of winter wheat (Figure 1): V1 genotype (model line of wasko.gl (−)—devoid of the main fractions of allergenic gluten proteins) and V2 genotype (model line of wasko.gl (+)—containing the full set of allergenic proteins).

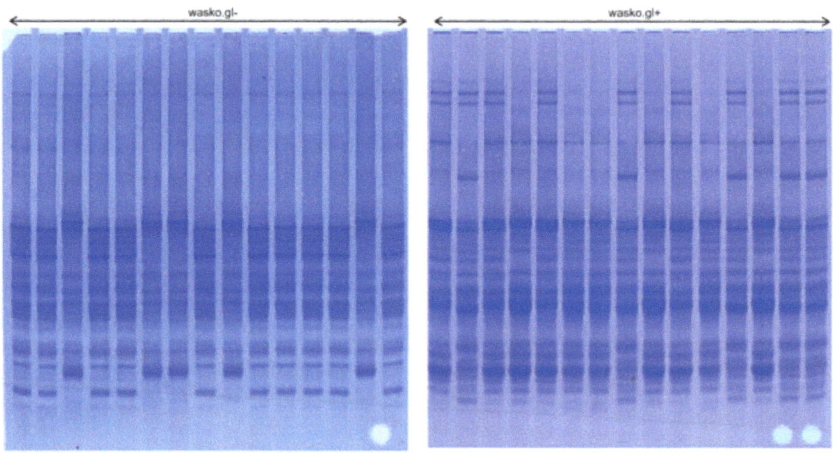

Figure 1. Genotype without (V1—wasko.gl−) and with (V2—wasko.pg+) ω—gliadin [9].

2. Materials and Methods

2.1. Experimental Conditions and Treatments

The study was conducted at the Prusy Experimental Station of the University of Agriculture in Krakow, located near Krakow (50°07′28″ N, 20°05′34″ E), Poland, during the 2019/2020 vegetation season. The research was conducted using the split-split-plot design with three replications. The first factor was the two genotypes of winter wheat specified as V1 (without allergenic protein) and V2 (with allergenic protein), and the second factor was three doses of sulfur fertilization (S dose) as S1—0, S2—20 and S3—40 kg S ha^{-1}. The third factor consisted of 6 doses of nitrogen fertilization (N dose) as N1—0, N2—40, N3—60, N4—80, N5—100 and N6—120 kg N ha^{-1}, respectively. The size of the small plots was 11.2 m^2. The experiment was established on chernozem. The soil was in the heavy category (36% fraction < 0.02 mm), with an acid reaction (pH$_{KCl}$ 6.2), low content of sulfur and medium content of available phosphorus (68 mg kg^{-1} DM), as well as of potassium (125 mg kg^{-1} DM).

2.2. Meteorological Conditions

The total precipitation during the growing season of winter wheat was 556.7 mm, and the average temperature for this period was 10.3 °C. In early spring (April), there were shortages of rainfall, which was manifested by yellowing of leaves in the shooting phase (BBCH 32), (Figure 2). This is a critical period in wheat development in terms of water needs. Water deficiency during this period can significantly reduce the level of yield, depending on the sensitivity of the genotype, and modify the efficiency of nitrogen utilization by plants. In May and June, optimum water requirement for wheat growth was observed, which positively influenced the inflorescence phase. Water shortage was observed at the early and late ripening phases, which had limited effect on yield formation.

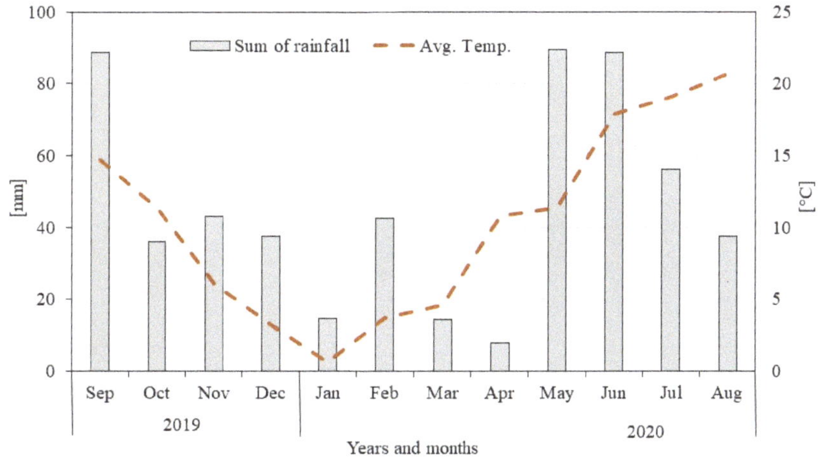

Figure 2. Meteorological conditions during vegetation of winter wheat.

2.3. UAV Remote Sensing

For the purposes of the experiment, a fixed-wing UAV platform was used, i.e., e-VTOL TRINITY F90+ (Quantum-Systems GmbH, Gliching, Germany), with a photogrammetric payload of double digital cameras: (i) high resolution RGB SONY UMC R10C 20.1 Mpx and (ii) multispectral RedEdge-M (MicaSense). Photos taken simultaneously in an automatic mission with these cameras (operator: ProGea SKY, Kraków, Poland) were designed with 75% side overlap and 75% frontal overlap, at the flight altitude of 75 m AGL. This allowed a field resolution of 2.0 cm for RGB orthoimagery (SONY Corporation, Tokio, Japan) and 5.2 cm GSD for 5-band imagery (MicaSense, Seattle, WA, USA) to be obtained. The spectral resolution of the RedEdge-M (MicaSense) camera is 5 spectral bands: RGB, RedEdge and NIR. The spectral specificity of the range of individual bands is presented in Table 1. The pilot performed an automatic mission on 22 May 2020 at peak sun hours in clear weather using two additional photos of dedicated calibration panel (MicaSense, Seattle, WA, USA), one before and one after the mission. Generation of orthoimages (*.TIFF; EPSG 2180) for 3-band RGB (SONY) and 5-band (RedEdge-M using calibration images) was performed using Metashape (Agisoft LCC, Petersburg, Russia) software (Figure 3). The geometric accuracy of the pixel position on the orthophotomap, according to the Meatashape report, about 3 cm (XY; RGB), was possible thanks to the use of the PPK (post processing kinematic) process using the local GNSS base station and 6 GCPs (Ground Control Points) measured with the RTK (Real Time Kinematic) GNSS method (Real Time Network; RMS 2.0 cm XYZ).

Table 1. Specification of wavelength and bandwidth of multispectral camera bands.

Band Name	Center Wavelength (nm)	Bandwidth FWHM (nm)
Blue	475	20
Green	560	20
Red	668	10
Red Edge	717	10
NIR	840	40

Figure 3. RGB UAV orthophoto (GSD 2.0 cm) based on SONY UMC R10C photos captured on 22 May 2020 (experiment area—magenta line). Almost no visible traces of lodging.

Based on the measurements performed in 5 spectral bands, the following vegetation indices were calculated based on equations (Table 2).

Table 2. Construction of selected indices analysis of winter wheat genotypes.

Canopy Index	Equation	Reference
Canopy Chlorophyll Content Index (CCCI)	CCCI = ((NIR − REDEDGE)/(NIR + REDEDGE))/((NIR − RED)/(NIR + RED)	Cammarano et al. [25]
Enhanced Vegetation Index (EVI)	EVI = 2.5 * ((NIR − RED)/(NIR + 6 × RED − 7.5 × BLUE + 1))	Matsushita et al. [26]
Green Normalized Difference Vegetation Index (GNDVI)	GNDVI = (NIR − GREEN)/(NIR + GREEN)	Chen et al. [27]
Normalized Difference Red Edge (NDRE)	NDRE = (NIR − REDEDGE)/(NIR + REDEDGE)	Thompson et al. [28]
Normalized Difference Vegetation Index (NDVI)	NDVI = (NIR − RED)/(NIR + RED)	Chen et al. [27]

2.4. Agrotechnical Details

The crop previous to the wheat was winter rape, followed by disking and plowing, harrowing and the cultivation of soil for sowing with an aggregate consisting of a cultivator and a string roller. Before establishing the experiment, phosphorus and potassium fertilizers were applied in amounts of 105 kg P_2O_5 and 100 kg K_2O ha^{-1} in the form of triple superphosphate and potassium salt and potassium sulfate (source of sulfur). Fertilization was applied in following pattern—S1: 184 kg potassium salt 60% + 263 kg superphosphate 40%; S2—111 kg potassium sulfate 50% + 91 kg potassium salt 60% + 263 kg superphosphate 40%; S3—222 kg potassium sulfate 50% + 263 kg superphosphate 40%.

The winter wheat was sown on 4 October 2019 in an amount of 350 grains per 1 m^2. The row spacing was 14 cm and the depth was 3 cm. The doses of nitrogen were applied in the following amounts and at the following times: N0—no nitrogen, N1—40 kg start vegetation (BBCH 25—6 March 2020), N2—60 kg (30 kg—beginning of spring vegetation (BBCH 25—6 March 2020) + 30 kg—shooting stage (BBCH 32—20 April 2020)), N3—80 kg (40 kg—(BBCH 25) + 40 kg—(BBCH 32)), N4—100 kg (50 kg—(BBCH 25) + 50 kg—(BBCH 32)), N5—120 kg (60 kg—(BBCH 25) + 60 kg—(BBCH 32)). Wheat was harvested with a plot combine machine during the full grain maturity stage (29 July 2020).

2.5. Vegetation Indices

LAI and NDVI indices were measured using the SunScan System with BF2 ground devices from Delta-T and an NTech Model 505 GreenSeeker HandHeld and calculated from the single image bands obtained by RedEdge-M (MicaSesne). Measurements were taken from the ground on 19 March 2020, 7 May, and 19 May 2020, and from UAV on 22 May 2020.

The geographic information system (GIS) raster layers representing vegetation indices selected for the study were generated and processed using ArcMap ArcGIS ver. 10.4. (Esri) software. The several GIS spatial analyses were performed to obtain the statistics (*Zonal statistics*) for every single investigation plot vectorized on high resolution 2 cm GSD RGB orthophoto. As the result the *.CSV file with basic statistics for every plot was exported.

2.6. Calculations and Statistical Analysis

The results for individual traits were statistically processed using the analysis of variance, simple correlation and curvilinear regression analysis (2nd-degree polynomials). The analysis of variance was performed for a 3-factor experiment in a split-split-plot design in 3 replications. The analysis was performed using Excel and Statistica software. Correlation analysis was performed for the mean values of individual indices and the grain yield. Based on the values for interaction of genotype x fertilization with N or S, the production functions for the grain yield were determined using the 2nd-degree polynomials separately for both genotypes.

Similar calculations were performed for individual vegetation indices, thus determining the level of nitrogen fertilization for the maximum value of the indices described by the square equation. The forecast of the production function and yield for both genotypes was calculated at the designated doses of nitrogen. The doses determined from the production function and from the function of the course of the trend for vegetation indices, as well as the expected yields, cancelled each other out and the dose reduction was calculated in the case of reaching the maximum value of the index and the corresponding level of yield reduction.

3. Results

3.1. Grain Yield and Agronomic Efficiency

Of the studied wheat genotypes, the grain yield was at the level of 4.62–6.01 Mg ha^{-1} for the genotype marked as V1, while for the V2 genotype the yield was 3.82–5.43 Mg ha^{-1} (Table 3). The differences between the yields of the genotypes were statistically significant, while there was no statistical differentiation in the traits for the other two factors (nitrogen and sulfur fertilization). However, with regard to nitrogen, there was a clear tendency to increase the yield, most often in the dose range of 0–80 kg ha^{-1} (Figure 4), while, in the case of sulfur fertilization, such a tendency was less visible (Figure 5). The agronomic efficiency ranged between 2 and 12.67 kg of grain per 1 kg nitrogen fertilization. The biggest value of this index was obtained at a dose of nitrogen of 60 kg N ha^{-1} for both genotypes (Figure 6).

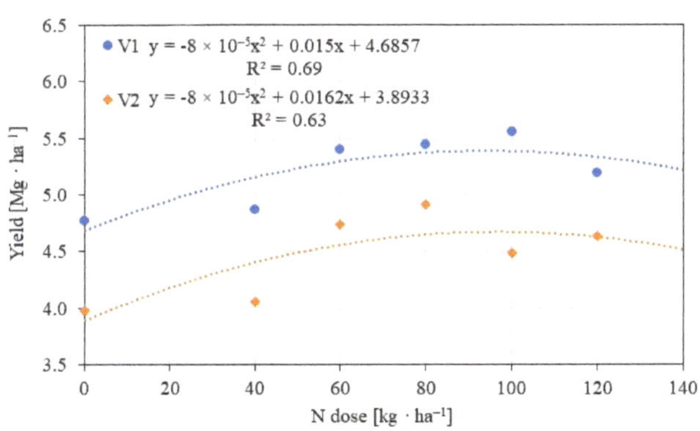

Figure 4. Grain yield of winter wheat genotypes depending on nitrogen fertilization.

Table 3. Grain yield (Mg ha^{-1}) of winter wheat genotypes under fertilization.

Fertilization (kg·ha^{-1})		Genotype		Mean
N Dose	**S Dose**	**V1**	**V2**	
0	0	4.62	3.88	4.25
	20	5.02	4.25	4.64
	40	4.69	3.82	4.26
Mean		4.78	3.98	4.38
40	0	4.84	4.15	4.50
	20	4.73	4.18	4.46
	40	5.04	3.85	4.45
Mean		4.87	4.06	4.47
60	0	5.30	4.76	5.03
	20	5.66	4.72	5.19
	40	5.24	4.74	4.99
Mean		5.40	4.74	5.07
80	0	5.26	5.43	5.35
	20	5.30	4.65	4.98
	40	5.77	4.67	5.22
Mean		5.45	4.91	5.18
100	0	5.67	3.99	4.83
	20	6.01	4.86	5.44
	40	5.01	4.61	4.81
Mean		5.56	4.49	5.03
120	0	4.72	4.84	4.78
	20	5.33	4.68	5.01
	40	5.52	4.38	4.95
Mean		5.19	4.47	4.83
Mean for genotype		5.21	4.47	4.84
Mean for S dose	0	5.07	4.51	4.79
	20	5.34	4.56	4.95
	40	5.21	4.34	4,78
$LSD_{p=0.05}$ for genotype		0.04		
$LSD_{p=0.05}$ for N dose		-		n.s. *
$LSD_{p=0.05}$ for S dose		-		n.s.

*—not significant. LSD—least significant difference. Genotype without (V1—wasko.gl−) and with (V2—wasko.pg+) ω—gliadin.

3.2. Vegetative Indices

The NDVI index before nitrogen application (BBCH 29) was on average at the level of 0.878 for the V1 genotype and 0.824 for the V2 genotype (Table 4). The significance of the differences in the dose of nitrogen fertilization was not confirmed statistically (Figure 7). The values of the NDVI index after applying a top dose of N is well presented (Figure 8). Both genotypes showed the same response to increased nitrogen doses as evidenced by the almost parallel course of lines determined by the regression function and coefficients of determination close to 1 (Figure 8). The mean value of the NDVI index during the heading period was 0.878 and 0.824 for genotypes V1 and V2, respectively. The difference between these means was statistically significant. The value of this index in the heading phase of these genotypes ranged from 0.792 to 0.870 for V2 and 0.584 to 0.902 for V1. Sulfur fertilization did not result in any significant differentiation in this trait, and neither did nitrogen fertilization; only in the latter case was there a clear tendency to increase the value. This index increased the value in the range of fertilization by 0–80 kg N ha^{-1} (Table 4).

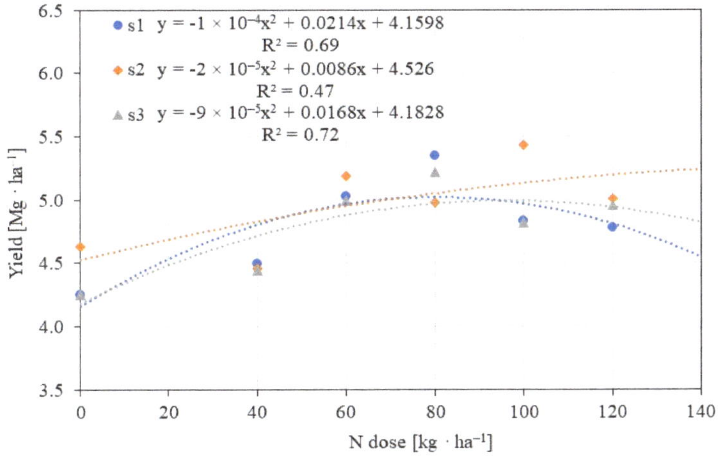

Figure 5. Grain yield of winter wheat genotypes depending on nitrogen and sulfur fertilization.

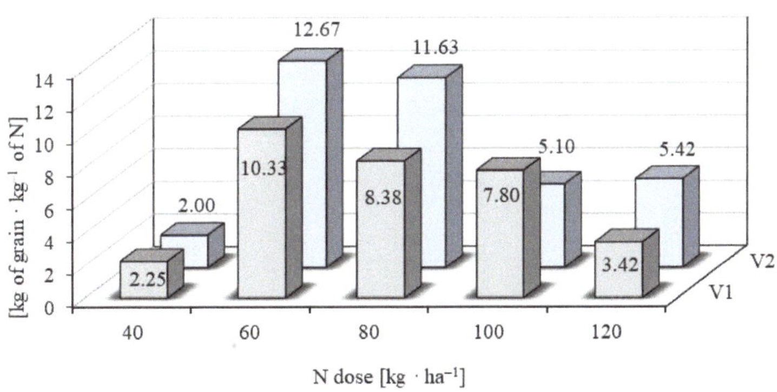

Figure 6. Agronomic efficiency of nitrogen doses.

Table 4. NDVI of winter wheat genotypes under fertilization.

Fertilization (kg·ha^{-1})		Genotype		Mean
N Dose	S Dose	V1	V2	
0	0	0.584	0.792	0.688
	20	0.877	0.823	0.850
	40	0.876	0.810	0.843
Mean		0.869	0.808	0.839
40	0	0.892	0.840	0.866
	20	0.893	0.800	0.847
	40	0.868	0.836	0.852

Table 4. *Cont.*

Fertilization (kg·ha^{-1})		Genotype		Mean
N Dose	S Dose	V1	V2	
Mean		0.884	0.825	0.855
60	0	0.864	0.870	0.867
	20	0.897	0.847	0.872
	40	0.866	0.824	0.845
Mean		0.876	0.847	0.862
80	0	0.902	0.832	0.867
	20	0.900	0.800	0.850
	40	0.901	0.851	0.876
Mean		0.901	0.828	0.865
100	0	0.886	0.815	0.851
	20	0.894	0.806	0.850
	40	0.869	0.854	0.862
Mean		0.883	0.825	0.854
120	0	0.861	0.821	0.841
	20	0.825	0.814	0.820
	40	0.871	0.800	0.836
Mean		0.853	0.812	0.833
Mean for genotype		0.878	0.824	0.851
Mean for S dose	0	0.876	0.828	0.852
	20	0.881	0.815	0.848
	40	0.875	0.829	0.852
$LSD_{p=0.05}$ for genotype		0.043		
$LSD_{p=0.05}$ for N dose		-		n.s.*
$LSD_{p=0.05}$ for S dose		-		n.s.

* not significant. LSD—least significant difference. Genotype without (V1—wasko.gl−) and with (V2—wasko.pg+) ω—gliadin.

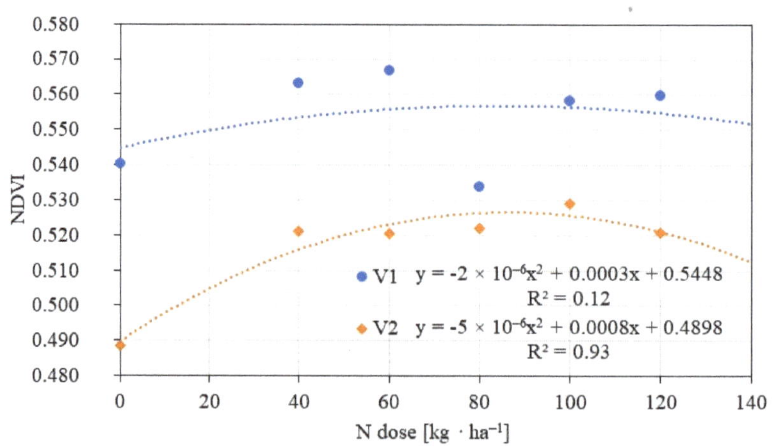

Figure 7. NDVI prior to application of nitrogen fertilization (19 March 2020)—BBCH 29.

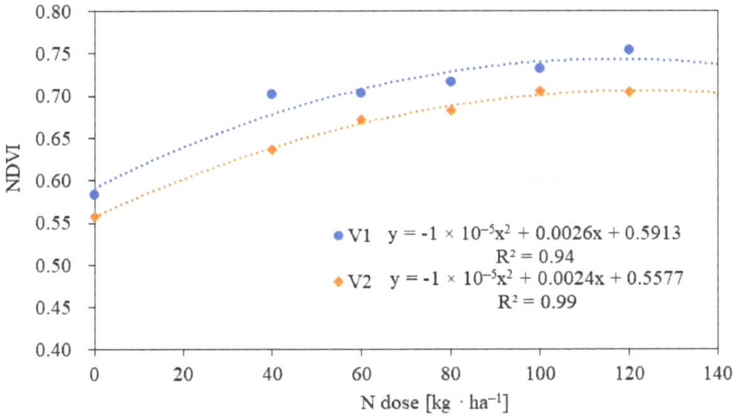

Figure 8. NDVI of winter wheat genotypes under N-fertilization (22 May 2020)—BBCH 39.

The LAI index was shaped according to the studied levels of sulfur and nitrogen fertilization in the range of 3.32–4.99 for genotype V1 and 2.83–4.41 for genotype V2 (Figure 9). Only the genetic factor had a significant influence on the value of this index. There were very small insignificant differences between the levels of sulfur fertilization, while nitrogen fertilization had a positive effect on increasing the value of this index with doses of over 120 kg N ha^{-1} for both genotypes. On average, within the range of 0–80 kg N ha^{-1}, the increase in the LAI index was from 4.86 to 5.25 (Figure 9). Higher doses caused lodging of plants, as shown in the photos from the experiment site (Figure 3).

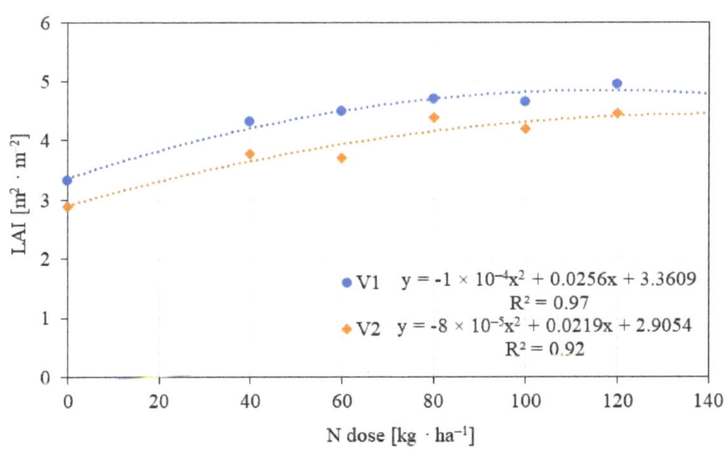

Figure 9. LAI of winter wheat genotypes under N-fertilization (22 May 2020)—BBCH 39.

The canopy chlorophyll content index (CCCI) index of the tested hypoallergenic wheat genotypes ranged from 0.572 to 0.646 for the V1 genotype and from 0.546 to 0.608 for the V2 genotype, respectively (Table 5). The differences between the cultivars were statistically insignificant but clearly marked. Sulfur fertilization did not significantly affect the value of this index. It seems that the optimal fertilization level is the dose of 20 kg S ha^{-1}. Nevertheless, nitrogen fertilization did significantly influence the value of this index (Figure 3). Nitrogen fertilization increased the index value with a dose of 80 kg N ha^{-1}, and then it decreased due to the lodging of plants and deterioration in growth conditions.

Table 5. CCCI of winter wheat genotypes under fertilization.

Fertilization (kg·ha^{-1})		Genotype		Mean
N Dose	S Dose	V1	V2	
0	0	0.572	0.554	0.563
	20	0.596	0.551	0.574
	40	0.581	0.554	0.568
Mean		0.583	0.553	0.568
40	0	0.632	0.588	0.610
	20	0.626	0.546	0.586
	40	0.606	0.580	0.593
Mean		0.621	0.571	0.596
60	0	0.597	0.608	0.603
	20	0.635	0.576	0.606
	40	0.616	0.557	0.587
Mean		0.616	0.580	0.598
80	0	0.645	0.590	0.618
	20	0.646	0.551	0.599
	40	0.632	0.582	0.607
Mean		0.641	0.574	0.608
100	0	0.614	0.568	0.591
	20	0.629	0.546	0.588
	40	0.617	0.587	0.602
Mean		0.620	0.567	0.594
120	0	0.614	0.596	0.605
	20	0.586	0.556	0.571
	40	0.585	0.547	0.566
Mean		0.595	0.566	0.581
Mean for genotype		0.613	0.569	0.591
Mean for S dose	0	0.612	0.584	0.598
	20	0.620	0.554	0.587
	40	0.606	0.568	0.587
$LSD_{p=0.05}$ for genotype		n.s.*		
$LSD_{p=0.05}$ for N dose		-		0.032
$LSD_{p=0.05}$ for S dose		-		n.s.

* not significant. LSD—least significant difference. Genotype without (V1—wasko.gl−) and with (V2—wasko.pg+) ω—gliadin.

Among the vegetation indices analyzed, the NDRE index showed the lowest values (Figure 3; Table 6). For the V1 genotype, they ranged from 0.489 to 0.582, while, for the V2 genotype, they ranged from 0.438 to 0.531. The differences in the values of this index for genotypes were statistically proven, as was the effect of nitrogen fertilization. The increase in the index value was recorded up to the dose of 80 kg N ha^{-1}. Sulfur fertilization caused insignificant reductions in the value of this index in relation to objects not fertilized with sulfur.

Table 6. NDRE of winter wheat genotypes under fertilization.

Fertilization (kg·ha^{-1})		Genotype		Mean
N Dose	S Dose	V1	V2	
0	0	0.489	0.439	0.464
	20	0.523	0.454	0.489
	40	0.509	0.449	0.479
Mean		0.507	0.447	0.477
40	0	0.564	0.496	0.530
	20	0.559	0.438	0.499
	40	0.527	0.487	0.507
Mean		0.550	0.474	0.512
60	0	0.518	0.531	0.525
	20	0.570	0.490	0.530
	40	0.536	0.459	0.498
Mean		0.541	0.493	0.517
80	0	0.582	0.493	0.538
	20	0.582	0.441	0.512
	40	0.570	0.496	0.533
Mean		0.578	0.477	0.528
100	0	0.545	0.463	0.504
	20	0.563	0.441	0.502
	40	0.539	0.503	0.521
Mean		0.549	0.469	0.509
120	0	0.534	0.473	0.504
	20	0.487	0.453	0.470
	40	0.540	0.438	0.489
Mean		0.520	0.445	0.483
Mean for cv.		0.541	0.469	0.505
Mean for S dose	0	0.539	0.483	0.511
	20	0.547	0.453	0.500
	40	0.537	0.472	0.505
$LSD_{p=0.05}$ for genotype		n.s.		
$LSD_{p=0.05}$ for N dose		-		0.047
$LSD_{p=0.05}$ for S dose		-		n.s.*

* not significant. LSD—least significant difference. Genotype without (V1—wasko.gl−) and with (V2—wasko.pg+) ω—gliadin.

The GNDVI index was slightly lower than the NDVI. The mean value of this index for the genotype V1 was 0.787 and for V2 was 0.736 (Table 7). In the case of this index, only the proven differences occurred, while no differences were found for sulfur and nitrogen fertilization. As in the case of NDVI, the value of the GNDVI also increased with an increase in the dose to 80 kg N ha^{-1} (V1) and up to 60 kg N ha^{-1} (V2).

Table 7. GNDVI of winter wheat genotypes under the fertilization.

Fertilization (kg·ha^{-1})		Genotype		Mean
N Dose	S Dose	V1	V2	
0	0	0.753	0.714	0.734
	20	0.778	0.726	0.752
	40	0.769	0.722	0.746
Mean		0.766	0.721	0.744
40	0	0.800	0.754	0.777
	20	0.799	0.714	0.757
	40	0.776	0.748	0.762
Mean		0.792	0.739	0.766
60	0	0.771	0.780	0.776
	20	0.807	0.751	0.779
	40	0.779	0.730	0.755
Mean		0.786	0.754	0.770
80	0	0.812	0.753	0.783
	20	0.814	0.717	0.766
	40	0.806	0.756	0.781
Mean		0.811	0.742	0.777
100	0	0.790	0.731	0.761
	20	0.801	0.717	0.759
	40	0.782	0.759	0.771
Mean		0.791	0.736	0.764
120	0	0.796	0.728	0.762
	20	0.747	0.725	0.736
	40	0.784	0.716	0.750
Mean		0.776	0.723	0.750
Mean for genotype		0.787	0.736	0.762
Mean for S dose	0	0.787	0.743	0.765
	20	0.791	0.725	0.758
	40	0.783	0.738	0.761
$LSD_{p=0.05}$ for genotype		0.043		
$LSD_{p=0.05}$ for N dose		-		n.s.*
$LSD_{p=0.05}$ for S dose		-		n.s.

* not significant. LSD—least significant difference. Genotype without (V1—wasko.gl−) and with (V2—wasko.pg+) ω—gliadin.

The last of the most important indices is the EVI index. Its values in this phase were in the range of 0.438 to 0.471, with mean values of 0.463 and 0.446 for genotypes V1 and V2, respectively. The trend in the nitrogen doses was similar to that for NDVI and GNDVI (Tables 4, 7 and 8).

Table 8. EVI index of winter wheat genotypes under fertilization.

Fertilization (kg·ha^{-1})		Genotype		Mean
N Dose	S Dose	V1	V2	
0	0	0.454	0.436	0.445
	20	0.462	0.446	0.454
	40	0.460	0.441	0.451
Mean		0.458	0.441	0.450
40	0	0.468	0.451	0.460
	20	0.468	0.438	0.453
	40	0.460	0.449	0.455
Mean		0.465	0.446	0.456
60	0	0.457	0.459	0.458
	20	0.469	0.452	0.461
	40	0.460	0.445	0.453
Mean		0.462	0.452	0.457
80	0	0.470	0.448	0.459
	20	0.470	0.437	0.454
	40	0.470	0.452	0.461
Mean		0.470	0.446	0.458
100	0	0.465	0.443	0.454
	20	0.468	0.440	0.454
	40	0.461	0.455	0.458
Mean		0.465	0.446	0.456
120	0	0.471	0.453	0.462
	20	0.447	0.442	0.445
	40	0.461	0.438	0.450
Mean		0.460	0.444	0.452
Mean for genotype		0.463	0.446	0.455
Mean for S dose	0	0.464	0.448	0.456
	20	0.464	0.442	0.453
	40	0.462	0.447	0.455
$LSD_{p=0.05}$ for genotype		0.043		
$LSD_{p=0.05}$ for N dose		-		n.s.*
$LSD_{p=0.05}$ for S dose		-		n.s.

* not significant. LSD—least significant difference. Genotype without (V1—wasko.gl−) and with (V2—wasko.pg+) ω—gliadin.

As shown by the correlation analysis, all indices from the low-ceiling level (using a drone) showed a high mutual correlation, which proves that they can be used as substitutes for each other. In addition, these indices (CCCI, NDRE, GNDVI, EVI and NDVI) showed a moderate correlation with wheat grain yield higher than the LAI index determined by ground measurement (Table 9). This could be due to the fact that the entire plot area was taken into account when measuring from the low-ceiling level and the LAI index was measured in four repetitions on the plot (Figure 10).

Table 9. Simple correlation coefficients between selected vegetation indices and the yield of hypoallergenic wheat grain (22 May 2020).

	CCCI	EVI	GNDVI	NDRE	NDVI	Yield
EVI	0.93	-				
GNDVI	0.93	0.84	-			
NDRE	0.98	0.96	0.94	-		
$NDVI^{-1}$	0.93	1.00	0.84	0.96	-	
Yield	0.68	0.66	0.61	0.66	0.66	-
LAI	0.48	0.42	0.49	0.46	0.41	0.59

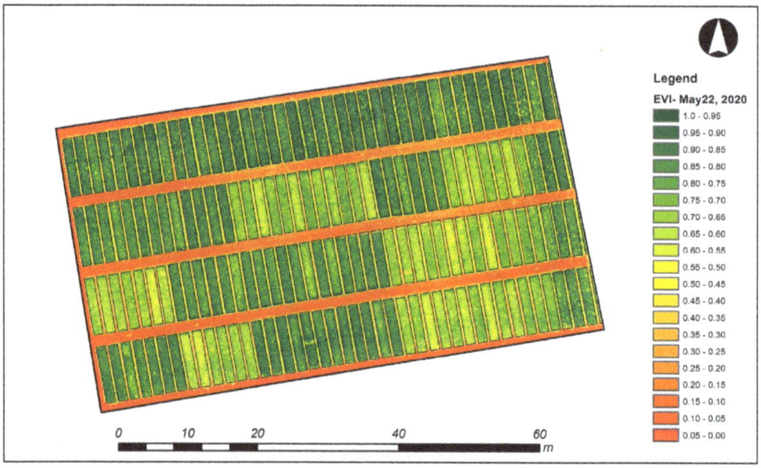

Figure 10. Visualization of EVI index depending on genotypes, sulfur and nitrogen fertilization.

Regression analysis was performed for the mean values of the interaction of genotypes and nitrogen fertilization, and, for the grain yield, the production function and the function for the vegetation indices, depending on the level of nitrogen fertilization, were determined. The doses determined from the production function were 94 and 101 kg N ha^{-1}, while the doses determined on the basis of the function for the vegetation indices were significantly lower, from 50 to 80 kg N ha^{-1} (Table 10). After substituting for the production function, the yields were lower by 10 to 150 kg of grain, which, in most cases except for the dose of 58 kg N ha^{-1}, eliminated the losses resulting from the yield reduction by reducing fertilization costs. The production function shows that lowering the dose to about 65 kg N ha^{-1} slightly decreased the level of wheat yield, especially that of the V2 genotype (Table 10). It should be emphasized that there was a significant reduction in GHG gas emissions as a result of limiting fertilization with the component with the greatest impact on the greenhouse effect (Figure 11).

Table 10. Comparison of maximum of nitrogen dose, as well as value of indices calculated for grain yield and vegetation indices of hypoallergenic genotypes of winter wheat.

Traits	Genotype	Equations: Coefficients of Equations $Y = Ax^2 + Bx + C$ Where x = Dose of N in kg	Equations: R^2	N kg per ha * (1)	Estimated Grain Yield ($Mg \cdot ha^{-1}$) (2)	Maximum Value of Indices Estimated From (3)	Decrease N Dose ($kg \cdot ha^{-1}$) (4)	In euro (5) (5) = 0.638 * (4)	Simulated Reduction in Grain Yield in kg (6)	Financial Loss after Yield Reduction (7) (7) = 0.21 * (6)	Result (8) (8) = (5) + (7)
Grain yield	v1	$Y = -0.00008x^2 + 0.015x + 4.69$	0.69	94	5.39	-	-	-	-	-	-
	v2	$Y = -0.00008x^2 + 0.0162x + 3.89$	0.63	101	4.71	-	-	-	-	-	-
LAI	v1	$Y = -0.0001x^2 + 0.0256x + 3.36$	0.97	128	5.30	5.56	−31.00	−19.65	−90	−18.90	−38.55
	v2	$Y = -0.00008x^2 + 0.0219x + 2.91$	0.92	137	4.60	4.91	−36.00	−22.92	−110	−23.10	−46.02
NDRE	v1	$Y = -0.00001x^2 + 0.0016x + 0.5046$	0.75	80	5.38	0.57	14.00	8.88	−10	−2.10	6.78
	v2	$Y = -0.000009x^2 + 0.0012x + 0.4466$	0.89	67	4.62	0.49	34.00	21.56	−90	−18.90	2.66
CCCI	v1	$Y = -0.00001\ x^2 + 0.0014\ x + 0.5814$	0.83	70	5.35	0.63	24.00	15.22	−40	−8.40	6.82
	v2	$Y = -0.000006x^2 + 0.0007x + 0.5535$	0.89	58	4.56	0.57	43.00	27.26	−150	−31.50	−4.24
NDVI	v1	$Y = -0.000007x^2 + 0.0008x + 0.8655$	0.58	64	5.32	0.89	30.00	19.02	−70	−14.70	4.32
	v2	$Y = -0.000007x^2 + 0.0009x + 0.8075$	0.78	57	4.60	0.84	37.00	23.46	−110	−23.10	0.36
GNDVI	v1	$Y = -0.000007x^2 + 0.001x + 0.7647$	0.71	71	5.35	0.80	23.00	14.58	−40	−8.40	6.18
	v2	$Y = -0.000007x^2 + 0.0009x + 0.7208$	0.89	64	4.60	0.75	37.00	23.46	−110	−23.10	0.36
EVI	v1	$Y = -0.000002x^2 + 0.0003x + 0.4578$	0.60	75	5.37	0.469	19.00	12.1	−20	−4.20	7.9
	v2	$Y = -0.000002x^2 + 0.0002x + 0.4411$	0.67	50	4.50	0.446	51.00	32.4	−210	−44.10	−11.7

* Solution of equations (First derivative): $2Ax + B = 0$, 1kg N = 0.634 Euro, 1kg grain = 0.21 Euro. Genotype without (V1—wasko.gl−) and with (V2—wasko.pg+) ω—gliadin.

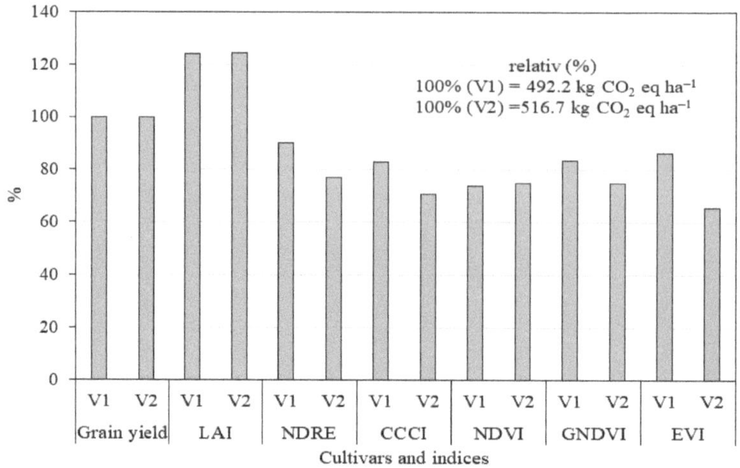

Figure 11. Simulation of reduction in GHG emissions from NPK fertilization depending on the index used in relation to the N dose resulting from the production function (grain yield). Genotype without (V1—wasko.gl−) and with (V2—wasko.pg+) ω—gliadin.

4. Discussion

4.1. Grain Yield and Fertilization Efficiency

The yields obtained in the experiment ranged from 3.61 to 6.01 Mg ha^{-1}, depending on the genotype and N and S fertilization. It should be noted that these genotypes are characterized by primary traits, e.g., difficult grain profitability. On the objects without fertilization, the yield was obtained at the level of 3.5 Mg ha^{-1}; therefore, assuming the maximum yield (approx. 6 Mg ha^{-1}), approximately 60 kg N ha^{-1} is optimal. Contemporary cultivars of common wheat give much better yields in Poland, as evidenced by the results obtained in the experiments conducted by Tabak et al. [29]. They showed that the optimal nitrogen dose was 217 kg N ha^{-1}, and the maximum yield was 8.251 Mg ha^{-1}. The production function depending on the nitrogen dose, calculated on the basis of yield, showed that the maximum yield should be expected at doses of 94 and 101 kg N ha^{-1} for genotypes V1 and V2, respectively, giving yields of 5.39 and 4.71 Mg ha^{-1}. This result is similar to those obtained in Salus model studies conducted by Basso et al. [30], in which they considered the dose of 90 kg N ha^{-1} to produce an economically viable crop. Zhang et al. [31], based on the linear plateau model, found that the optimal N dose (for field trials conducted at 120 sites) varied from 84 kg to 270 kg N ha^{-1}, with a mean value of 138 kg ha^{-1}, under which the maximum wheat yield varied from 5213 Mg ha^{-1} to 8785 Mg ha^{-1}. Nitrogen is an essential ingredient for the realization of the potential production capacity of the varieties. Doses at a level of 120 kg ha^{-1} should ensure a yield of 8 Mg ha^{-1}. Excess ingredient causes higher uptake and increased washout. The agronomic efficiency of nitrogen fertilization of the studied genotypes was low, in the range of 2–12.67 kg of grain per kg of nitrogen. According to Dobermann [32], the value of this indicator is usually in the range from 10 to 30 kg of grain per 1 kg of N, and under conditions of high nitrogen deficit or under favorable conditions of vegetation, these values exceed 25 kg of grain per 1 kg of N. According to Kołodziejczyk [23] the N agronomic efficiency (NAE) was significantly influenced by the weather conditions, level of nitrogen fertilization and interactions between these factors, but also by the spring wheat cultivars. Depending on the level of N fertilization and the year of study, the NAE ranged from 4.7 to 43.4 kg kg^{-1}. NAE was significantly higher in years with lower amounts of rainfall. The highest N agronomic efficiency, of 32.7 kg kg^{-1}, was observed for the dose of 60 kg N ha^{-1}. Increasing the nitrogen dose to 120 and to 150 kg N ha^{-1} resulted in a decrease in NAE by 39 and 54%. López-Bellido and López-Bellido [33] observed the nitrogen fertilization of winter wheat in doses from 50 to 150 kg N ha^{-1} to have a significant influence on NAE values, which ranged from 4.9 to 7.2 kg kg^{-1}. NAE values which were several times lower were also confirmed in studies by Delogu et al. [22]. The N agronomic efficiency was similar in barley and wheat (8.7 and 9.2 kg kg^{-1} of N applied, respectively), suggesting that both species respond equally to nitrogen fertilization. Nevertheless, due to the lower nitrogen use efficiency (NUtE) value, wheat requires high nitrogen fertilization to optimize yields, while, in barley, the lower nitrogen level necessary to obtain the highest yields allows this crop to perform better under conditions of low application inputs [33].

4.2. Physiological and Vegetative Indices

The LAI index is a physiological and measurable index, as it describes the size of the assimilation area of leaves per unit area of land [18]. The remaining indicators are dimensionless values of various ranges, well correlated with the yield, aboveground biomass and the LAI index [17,34,35]. The low value of the LAI index translates into the amount of photosynthesis and, consequently, the amount of biomass, including the useful yield. The LAI index determined at the BBCH 39 phase showed varietal differentiation. The better yielding genotype V1 was characterized by an 8% larger assimilation area compared to V2. A quadratic function plotted on the basis of the LAI vs. the dose of N showed that this index would reach the maximum value at doses of 128 and 137 kg N ha^{-1}, which, determined from the production function, corresponds to the yield level of 5.34 and 4.62 Mg ha^{-1}. These values are slightly lower than those determined from the production function, and the nitrogen doses are almost 30 kg higher. An excessive value of the LAI index is not beneficial for yielding, as it reduces the use of PAR due to mutual shading of leaves and lodging of the canopy, and worsens the conditions for plant growth and development. An ideal canopy has an area of 3–5 m^2 of leaves per 1 m^2 of soil, depending on the angle of the leaves. There is an interdependence between the studied indicators, meaning that they can be substituted for each other. For example, the most common relationship between LAI and NDVI and GNDVI is exponential, while between LAI and CCCI, NDRE and RVI, the relationship is rectilinear [25]. The EVI indicator is lower: on average, when the NDVI value is 0.9, the EVI value is 0.75. Depending on the density of the canopy, its value is 50–83% of the NDVI index. NDVI is a popular vegetation index, but it does not show the best correlation with yield and biomass for all species and development stages. In the present study, the NDVI index, in the initial period of spring growth (BBCH 29), showed insignificant and undirected random variation and ranged from 0.515 to 0.550. In the second period of measurement, there was shown to be significant differentiation between genotypes and a tendency described by the square function for both genotypes, depending on the level of nitrogen fertilization. The values of this index ranged from 0.792 to 0.902. The quadratic function of the progress of this indicator allows its peak to be determined, which both genotypes reached at 64 kg N ha^{-1}. At this dose of N, the yield was at the level of 5.32 and 4.0 Mg ha^{-1} for genotypes V1 and V2, respectively. According to Fu et al. [36], in late stages of development, the NDRE index is a better estimator. The flowering phase [37] was the best stage (phase) of development for prediction based on models based on this indicator. The determination coefficients between the vegetation index and yield for NDVI at the heading stage were 0.59–0.76 while, for NDRE at the flowering stage, they were 0.69–0.78.

4.3. Simulated Reduction in GHG Emissions

Crop production is a significant contributor to total anthropogenic greenhouse gas emissions into the atmosphere. Management practices involving soil tillage, sowing, fertilizer application, irrigation and pest management have a significant influence on emissions of carbon dioxide (CO_2) and nitrous oxide (N_2O). Per 1 kg of conventionally grown wheat grain, the individual emission components are: agrotechnical operations 0.078, fertilizers 0.221, pesticides 0.001, seeds 0.023 and field emission of 0.137 CO_2 eq. per 1 kg of grain field [38] According to Kumar et al. [39], CO_2 emission per ha in wheat fertilized with different N-doses (0–240 kg N ha^{-1}) ranged from 292.3 to 765.3 kg CO_2. The application of nitrogen at a dose of 150 kg ha $^{-1}$ had the highest GHG emission per ha—1974.1 kg CO_2, compared to the control.

All the agricultural inputs (as already mentioned) bring an environmental impact, which can be quantified by the life cycle assessment method [38]. Moreover, the impact on the environment during the cultivation of cereals is most evident in the input of fertilizers, especially nitrogen fertilizers [23]. However, for example, Bernas et al. [39] indicated that intensive cultivation practices, i.e., practices with high fertilizer inputs, do not necessarily confer the highest environmental impact. In this respect, the achieved yield level, the choice of allocation approach and the functional unit play a dominant role.

5. Conclusions

The yields obtained in the experiment varied depending on the genotype. The production function depending on the nitrogen dose showed that the maximum yield should be expected at doses of 94 and 101 kg N ha^{-1} for genotypes V1 (without allergenic protein) and V2 (with allergic protein), respectively, giving yields of 5.39 and 4.71 Mg ha^{-1}. The LAI index reached its maximum values at doses of about 30 kg higher (128 and 137 kg N ha^{-1}). On the basis of the values of the tested vegetation indices, the highest doses of N should be applied using the NDRE index, and the lowest ones based on the EVI index, and, in the latter case, a reduction in yield of more than 0.2 Mg ha^{-1} in the V2 genotype should be taken into account. Preliminary estimates indicate that all vegetation indices except LAI show lower nitrogen doses than those determined from the LAI function and will not differ significantly from the dose resulting from the production function. The simple correlation analysis proved that the vegetation indices based on radiometric measurements show a moderate correlation with the grain yield, but the difference in the forecasts are insignificant. The study indicates a clear need to continue field studies dedicated to the monitoring and assessment of hypoallergenic wheat productivity using UAV remote sensing techniques.

Author Contributions: Conceptualization, B.K.; M.R. and J.W.; methodology, B.K., A.S., J.W., M.R, P.W. and G.P.; software, A.O. and W.G.; validation, A.O., R.W., P.W. and B.K.; formal analysis, R.W., B.K. and W.G.; investigation, M.K., A.K.-K., B.K., A.O. and R.W.; resources, B.K., P.W. and A.O.; data curation, B.K., P.W. and M.R.; writing—original draft preparation, B.K., M.R., A.O.,M.K.,P.W.,A.K.-K., R.W.,A.S.,G.P.,W.G.; writing—review and editing, B.K., R.W., P.W., A.K.-K., A.O., W.G., G.P., A.S., J.W., M.K. and P.W.; visualization, P.W., A.O., R.W. and A.K.-K.; supervision, B.K., A.O. and M.R.; project administration, B.K., J.W. and M.R.; funding acquisition, B.K. All authors have read and agreed to the published version of the manuscript.

Funding: This research was funded by The Polish National Centre for Research and Development, Grant POIR.04.01.04-00-0051/18-00, acronym HYPFLO.

Institutional Review Board Statement: Not applicable.

Data Availability Statement: Not applicable.

Conflicts of Interest: The authors declare no conflict of interest.

References

1. Wrigley, C.W.; Bekes, F.; Bushuk, W. Gluten: A Balance of Gliadin and Glutenin. In *Gliadin and Glutenin—The Unique Balance of Wheat Quality*; Wrigley, C., Bekes, F., Bushuk, W., Eds.; AACC International: St. Paul, MN, USA, 2006; pp. 3–32.
2. Wrigley, C.W.; Bietz, J.A. Proteins and Amino Acids. In *Wheat: Chemistry and Technology*; Pomeranz, Y., Ed.; AACC: St. Paul, MN, USA, 1988; pp. 159–275.
3. Seilmeier, W.; Belitz, H.D.; Wieser, H. Separation and quantitative determination of high-molecular-weight subunits of glutenin from different wheat varieties and genetic variants of the variety Sicco. *Z. Lebensm. Unters. Forsch.* **1991**, *192*, 124–129. [CrossRef]
4. Baldacci, S.; Maio, S.; Cerrai, S.; Sarno, G.; Baiz, N.; Simoni, M.; Annesi-Maesano, I.; Viegi, G. Allergy and asthma: Effects of the exposure to particulate matter and biological allergens. *Respir. Med.* **2015**, *109*, 1089–1104. [CrossRef] [PubMed]

5. Sapone, A.; Bai, J.C.; Ciacci, C.; Dolinsek, J.; Green, P.H.; Hadjivassiliou, M.; Kaukinen, K.; Rostami, K.; Sanders, D.S.; Schumann, M.; et al. Spectrum of gluten-related disorders: Consensus on new nomenclature and classification. *BMC Med.* **2012**, *10*, 13. Available online: https://www.biomedcentral.com/1741-7015/10/13/-ins6PHR (accessed on 1 January 2020). [CrossRef] [PubMed]
6. Palosuo, K.; Varionen, E.; Kekki, O.M. Wheat ω-5 gliadin is a major allergen in children with immediate allergy to ingested wheat. *J. Allergy Clin. Immunol.* **2001**, *108*, 634–638. [CrossRef] [PubMed]
7. Morita, E.; Matsuo, H.; Mihara, S. Fast ω-gliadin is a major allergen in wheat-dependent exercise-induced anaphylaxis. *J. Dermatol. Sci.* **2003**, *33*, 99–104. [CrossRef]
8. Waga, J.; Skoczowski, A. Development and characteristics of ω-gliadin-free wheat genotypes. *Euphytica* **2014**, *195*, 105–116. [CrossRef]
9. Stawoska, I.; Waga, J.; Wesełucha-Birczynska, A.; Dziurka, M.; Podolska, G.; Aleksandrowicz, E.; Skoczowski, A. Does Nitrogen Fertilization Affect the Secondary Structures of Gliadin Proteins in Hypoallergenic Wheat? *Molecules* **2022**, *27*, 5684. [CrossRef]
10. Geipel, J.; Link, J.; Wirwahn, J.A.; Claupein, W. A Programmable Aerial Multispectral Camera System for In-Season Crop Biomass and Nitrogen Content Estimation. *Agriculture* **2016**, *6*, 4. [CrossRef]
11. Li, M.; Shamshiri, R.R.; Weltzien, C.; Schirrmann, M. Crop Monitoring Using Sentinel-2 and UAV Multispectral Imagery: A Comparison Case Study in Northeastern Germany. *Remote Sens.* **2022**, *14*, 4426. [CrossRef]
12. Walsh, O.S.; Shafian, S.; Marshall, J.M.; Jackson, C.; McClintick-Chess, J.R.; Blanscet, S.M.; Swoboda, K.; Thompson, C.; Belmont, K.M.; Walsh, W.L. Assessment of UAV Based Vegetation Indices for Nitrogen Concentration Estimation in Spring Wheat. *Adv. Remote Sens.* **2018**, *7*, 71–90. [CrossRef]
13. Huete, A.; Justice, C. MODIS Vegetation Index (MOD 13) Algorithm Theoretical Basis Document, 1999. Available online: https://modis.gsfc.nasa.gov/data/atbd/atbd_mod13.pdf (accessed on 1 January 2020).
14. Boiarskii, B.; Hasegawa, H. Comparison of NDVI and NDRE Indices to Detect Differences in Vegetation and Chlorophyll Content International Conference on Applied Science, Technology and Engineering. *J. Mech. Cont. Math. Sci.* **2019**, *4*, 20–29.
15. Huang, S.; Tang, L.; Hupy, J.P.; Wang, Y.; Shao, G. Commentary review on the use of normalized diference vegetation index (NDVI) in the era of popular remote sensing. *J. For. Res.* **2021**, *32*, 1–6. [CrossRef]
16. Bausch, W.C.; Duke, H.R. Remote sensing of plant nitrogen status in corn. *Trans. Am. Soc. Agric. Eng.* **1996**, *39*, 1869–1875. [CrossRef]
17. Carlson, T.N.; Ripley, D.A. On the relation between NDVI, fractional vegetation cover, and leaf area index. *Remote Sens. Environ.* **1997**, *62*, 241–252. [CrossRef]
18. Fang, H.; Baret, F.; Plummer, S.; Schaepman-Strub, G. An overview of global leaf area index (LAI): Methods, products, validation, and applications. *Rev. Geophys.* **2019**, *57*, 739–799. [CrossRef]
19. Węgrzyn, A.; Klimek-Kopyra, A.; Dacewicz, E.; Skowera, B.; Grygierzec, W.; Kulig, B.; Flis-Olszewska, E. Effect of Selected Meteorological Factors on the Growth Rate and Seed Yield of Winter Wheat—A Case Study. *Agronomy* **2022**, *12*, 2924. [CrossRef]
20. Adhikari, C.; Bronson, K.F.; Panuallah, G.M.; Regmi, A.P.; Saha, P.K.; Dobermann, A.; Olk, D.C.; Hobbs, P.R.; Pasuquin, E. On-farm soil N supply and N nutrition in the rice-wheat system of Nepal and Bangladesh. *Field Crop. Res.* **1999**, *64*, 273–286. [CrossRef]
21. Diacono, M.; Rubino, P.; Montemurro, F. Precision nitrogen management of wheat. A review. *Agron. Sustain. Develop.* **2013**, *33*, 219–241. [CrossRef]
22. Delogu, G.; Cattivelli, L.; Pecchioni, N.; De Falcis, D.; Maggiore, T.; Stanca, A.M. Uptake and agronomic efficiency of nitrogen in winter barley and winter wheat. *Eur. J. Agron.* **1998**, *9*, 11–20. [CrossRef]
23. Kołodziejczyk, M.; Kulig, B.; Oleksy, A.; Szmigiel, A. The effectiveness of N-fertilization and microbial preparation on spring wheat. *Plant Soil. Environ.* **2013**, *59*, 335–341. [CrossRef]
24. Grohs, D.S.; Bredemeier, C.; Mundstock, C.M.; Poletto, N. Model for yield potential estimation in wheat and barley using the Greenseeker sensor. *Engenharia Agrícola.* **2009**, *29*, 101–112. [CrossRef]
25. Cammarano, D.; Fitzgerald, G.; Basso, B.; O'Leary, G.; Chen, D.; Grace, P.; Fiorentino, C. Use of the Canopy Chlorophyl Content Index (CCCI) for Remote Estimation of Wheat Nitrogen Content in Rainfed. *Environ. Agron. J.* **2011**, *103*, 1597–1603. [CrossRef]
26. Matsushita, B.; Yang, W.; Chen, J.; Onda, Y.; Qiu, G. Sensitivity of the vegetation index (EVI) and normalized difference vegetation index (NDVI) to Topographic effects: A case study in high-density cypress forest. *Sensors* **2007**, *7*, 2636–2651. [CrossRef] [PubMed]
27. Chen, A.; Orlov-Levin, V.; Meron, M. Applying high-resolution visible-channel aerial imaging of crop canopy to precision irrigation management. *Agric. Water Manag.* **2019**, *216*, 196–205. [CrossRef]
28. Thompson, C.; Guo, W.; Sharma, B.; Ritchie, G.L. Using normalized difference red edge index to assess maturity in cotton. *Crop. Sci.* **2019**, *59*, 2167–2177. [CrossRef]
29. Tabak, M.; Lepiarczyk, A.; Filipek-Mazur, B.; Lisowska, A. Efficiency of Nitrogen Fertilization of Winter Wheat Depending on Sulfur Fertilization. *Agronomy* **2020**, *10*, 1304. [CrossRef]
30. Zhang, S.; Gao, P.; Tong, Y.; Norse, D.; Lu, Y.; Powlson, D. Overcoming nitrogen fertilizer over-use through technical and advisory approaches: A case study from Shaanxi Province, northwest China. *Agric. Ecosyst. Environ.* **2015**, *209*, 89–99. [CrossRef]
31. Dobermann, A. *Nutrient Use Efficiency: Measurement and Management*; IFA International Workshop on Fertilizer Best Management Practices: Brussels, Belgium, 2007; pp. 1–28.

32. López-Bellido, R.J.; López-Bellido, L. Efficiency of nitrogen in wheat under Mediterranean condition: Effect of tillage, crop rotation and N fertilization. *Field Crop. Res.* **2001**, *71*, 31–46. [CrossRef]
33. Lopez-Bellido, L.; Lopez-Bellido, R.J.; Redondo, R. Nitrogen efficiency in wheat under rainfed Mediterranean conditions as affected by split nitrogen application. *Field Crop. Res.* **2005**, *94*, 86–97. [CrossRef]
34. Prabhakara, K.; Hively, W.D.; McCartych, G.W. Evaluating the relationship between biomass, percent groundcover and remote sensing indices across six winter cover crop. *Int. J. Appl. Earth Observ. Geoinformat.* **2015**, *39*, 88–102, Fu, Z.; Jiang, J.; Gao, Y.; Krienke, B.; Wang, M.; Zhong, K.; Cao, Q.; Tian, Y.; Zhu, Y.; Cao, W.; Liu, X. Wheat Growth Monitoring and Yield Estimation based on Multi-Rotor Unmanned Aerial Vehicle. *Remote Sens.* **2020**, *12*, 508. [CrossRef]
35. Zhao, Y.; Potgieter, A.B.; Zhang, M.; Wu, B.; Hammer, G.L. Predicting Wheat Yield at the Field Scale by Combining High-Resolution Sentinel-2 Satellite Imagery and Crop Modelling. *Remote Sens.* **2020**, *12*, 1024. [CrossRef]
36. Jelínková, Z.; Moudrý, J.; Bernas, J.; Kopecký, M.; Moundry, J.; Konvalina, P. Environmental and economic aspects of Triticum aestivum L. and Avena sativa growing. *Open Life Sci.* **2016**, *11*, 533–541. [CrossRef]
37. Kumar, R.; Karmakar, S.; Minz, A.; Singh, J.; Kumar, A.; Kumar, A. Assessment of Greenhouse Gases Emission in Maize-Wheat Cropping System Under Varied N Fertilizer Application Using Cool Farm Tool. *Front. Environ. Sci.* **2021**, *9*, 355. [CrossRef]
38. Dijkman, T.J.; Basset-Mens, C.; Antón, A.; Núñez, M. LCA of Food and Agriculture. In *Life Cycle Assessment*; Hauschild, M.Z., Rosenbaum, R.K., Olsen, S.I., Eds.; Springer: Cham, Switzerland, 2018; pp. 723–754, ISBN 978-3-319-56474-6.
39. Bernas, J.; Bernasová, T.; Kaul, H.P.; Wagentristl, H.; Moitzi, G.; Neugschwandtner, R.W. Sustainability Estimation of Oat: Pea Intercrops from the Agricultural Life Cycle Assessment Perspective. *Agronomy* **2021**, *11*, 2433. [CrossRef]

agriculture

MDPI

Article

Evaluation of the Quality and Possible Use of a New Generation of Agricultural Nets for Packing Bulk Materials in Terms of the Aspect of Reducing the Environmental Burden

Ireneusz Kowalik [1,*], Bogna Zawieja [2], Piotr Rybacki [1] and Krzysztof Krzyżaniak [1]

[1] Department of Agronomy, Faculty of Agronomy, Horticulture and Bioengineering, Poznań University of Life Sciences, Dojazd 11, 60-632 Poznań, Poland

[2] Department of Mathematical and Statistical Methods, Faculty of Agronomy, Horticulture and Bioengineering, Poznań University of Life Sciences, Wojska Polskiego 28, 60-637 Poznań, Poland

* Correspondence: ireneusz.kowalik@up.poznan.pl

Abstract: In modern agriculture, packaging materials are becoming an important means of production in the technologies for harvesting bulk materials. The agricultural net currently used for this purpose is usually made of HDPE—high-density polyethylene. The aim of the study was to evaluate the agricultural net produced in light technology under the commercial name of Covernet. Based on the tests conducted for nine variants of different models of round balers and different bulk materials collected by them, it can be concluded that, in each case, the net (Tama LT) wrapped the cylindrical bales well or very well. The mean elongation of COVERNET during bale wrapping was over 8% for the tested machines and harvested materials. The tests confirmed the usefulness of the new generation of agricultural nets (Tama LT) for wrapping various agricultural bulk materials of various humidities. There is an urgent need to develop and implement in practice a technology for recovering used agricultural nets and converting them into granules that can be used again in their production.

Keywords: sustainable processing practices; net produced in light technology; agricultural net; packing agricultural bulk materials

1. Introduction

The intensive development of agriculture is connected with the biological progress and the breeding of modern variations of crop species that are more efficient and better adapted to local environmental and climatic conditions [1–7]. Another aspect of this progress is the implementation of new production technologies and the adaptation of new generations of materials for agricultural production which make farming more sustainable [8–12]. In many cases, measures taken in this area also enable the reduction in the negative impact of agriculture on directly related ecosystems. The consumption of materials that are harmful to the environment is reduced when harvesting straw and fodder [13]. In modern agriculture, packaging materials are becoming an important means of production in the technologies for harvesting bulk materials. The main problem with this method of packaging bulk materials is the generation of thousands of metric tons of waste. Another adverse effect during the HDPE production is the emission of greenhouse gases into the atmosphere. The carbon footprint of HDPE production is 1.60 kg $CO_2 \cdot kg^{-1}$ of polyethylene granules [14]. The agricultural net currently used for this purpose is usually made of HDPE—high-density polyethylene [15–20]. Manufacturers use various types of HDPE in the net production process [21]. High-grade raw materials of better quality make it possible to produce polymers of greater strength. In contrast, lower-quality raw materials are not as durable when using material of the same weight [22–24]. Therefore, manufacturers utilize thicker base material to improve the strength of agricultural nets. This means more of a consumption of raw material and more HDPE required per 1 running meter of net.

This leads to an increase in the weight of the net roll and its diameter while maintaining the same length of the net. As a result, it is impossible to produce a net from a standard raw material of higher strength without increasing the diameter and weight of the roll. Considering this situation and new requirements for environmental protection, the Tama Plastic Industry in Poland has introduced an innovative solution for net production, called Tama Light Technology (Tama LT). An advanced HDPE formula was used for this purpose in combination with a completely new production technology. The technology called (Tama LT) has allowed us to manufacture a net that is lighter and, at the same time, much more durable compared to the standard net. The introduction of a new generation of lightweight agricultural netting on the market required the verification of its suitability for packing agricultural bulky materials using baling presses.

2. Materials and Methods

2.1. Research Subject

Rolls of the Tama LT net under the trade name COVERNET were used in the field tests. COVERNET is a new-generation product for packing agricultural bulk materials with a minimum breaking strength of 2.5 kN. The aquamarine color net was 123 cm wide, with a guaranteed length of 2000 m and a gross roll weight of 20.50 kg.

2.2. Test Conditions

Before starting the operation of each round baler and the harvesting of bulk materials, it was checked whether the machine settings were in accordance with the manufacturer's recommendations. It was also checked if the owner did not make any structural changes to the machine. Then, the agrotechnical conditions during field tests were determined: the place of the test, the characteristics of the machine unit, the weather and agrotechnical conditions and the type of harvested plant material. The air temperature was measured with the LB-531 thermohygrometer (LAB-EL Laboratory Electronics, Herbaciana, Reguły, Poland) with an accuracy of 0.01 °C.

2.3. Measurement Methods

Before putting a roll of COVERNET in the machine's hopper, the distances between the net threads were measured to compare them after the bale had been wrapped. Comparing the distance between the threads in the net roll and in the wrapped bale allowed for determining the percentage net elongation during wrapping. For this purpose, the base of "10 triangles in the net" extended from a roll was calculated on a level surface before being placed in the machine's hopper, and then their length was measured. The measurement was conducted in four repetitions, and the measured distances of "10 bases of triangles" were recorded. Once the bale was wrapped, the base of the "10 triangles in the net" was calculated again on the inner layer of the net surface, and their length was measured. The outer net layer was not included in the measurements. The number of net wrap layers was determined (NL). Once the bale was unloaded from the round baler, the width of the net coverage was measured from the front and back of the bale. The measurement was conducted with a tape measure in four repetitions, and the measured distances were recorded with an accuracy of 1 mm.

2.4. Statistical Analysis

Parametric and non-parametric methods were applied in the statistical analysis of the measurement results, depending on the results of tests verifying the analysis assumptions. The distribution normality was checked using the Shapiro–Wilk test. Then, Pearson's correlation or non-parametric Spearman's rank correlation was used to determine whether the results of the measurements of net features depend on the order in which the bales were formed. The correlation between length 10Δ [cm] (LEN 10Δ) and the percentage net elongation (ELO) were also checked. The t-distribution method or the Wilcoxon non-parametric method (when the distribution normality assumption was not met) was used to compare the averages of two dependent populations. These methods were used to compare the mean net coverage at the bale front (CF) and at the bale back (CB). A correlation analysis between features of different types and a comparison of averages between features of the same type were carried out to determine whether all features should be analyzed separately. If no significant differences between the averages or a strong correlation was shown, only selected features were analyzed. The conclusions on these features were transferred to the other corresponding features: length 10Δ and elongation (ELO). In order to visually assess the results of the tests of round balers and the bulk materials collected by them, descriptive statistics were calculated and presented in a graphic form as violin plots. These charts showed the range of feature variability and the number of observations corresponding to a given feature value. A variance analysis was carried out for each of the features, with round balers and collected bulk materials as factors. Before starting the analysis, the homogeneity of variance of the compared factor levels was checked using Levene's test. If the assumptions of the normal variance analysis were not met, the non-parametric Kruskal–Wallis test was used. All analyses were carried out in the STATISTICA statistical package, and the violin plots were made using the ggplot2 procedure on the R statistical platform.

3. Results

The COVERNET from Tama, produced in light technology (Tama LT), was evaluated during field tests in the agrotechnical period between 3 June 2017 and 25 August 2017 during the harvesting of various agricultural bulk materials with round balers. The tests were conducted in several regions of Poland in farms equipped with a round baler, which voluntarily agreed to test the new generation of agricultural nets. The following bulk materials were collected: dried mixed-grass fodder intended for silage and straw from various cereal species: winter rye, winter triticale, winter wheat and spring barley. The air temperature during the harvesting of dried greens fodders ranged from 19 to 27 °C. During the harvesting of straw from various species of cereals, the air temperature ranged from 21 to 26 °C. COVERNET was tested in round balers from domestic manufacturers of agricultural machinery: Sipma S.A., UNIA Sp. z.o.o., POL-MOT Warfama S.A. and METAL-FACH Sp. z o.o. (Table 1).

Table 1. Variants harvest of bulk materials during tests of the COVERNET mesh produced in the Tama LT technology.

No	Round Baler Type	Bulk Material				
		Dried GreenFodder	**Spring Barley Straw**	**Winter Triticale Straw**	**Winter Wheat Straw**	**Winter Rye Straw**
1.	Sipma Z-569/1 Farna II	+	+	-	-	-
2.	UNIA DF 1,7 Zd	+	-	-	-	-
3.	UNIA Df 1,8 Dd	+	-	-	-	-
4.	Sipma PS 1211 Farma PLUS	-	-	+	-	-
5.	Warfama Z-543	-	+	-	+	-
6.	Metal-Fach Z-562	-	-	-	+	+

The collected test results were evaluated with the Shapiro–Wilk test to verify the null hypothesis confirming their normal distribution. The test showed that most of the analyzed round balers and the bulk materials they collected did not meet this assumption (Table 2).

Table 2. *p*-values of the Shapiro–Wilk test of normality.

Trait of Net	Total	Harvesting Variant *								
		1.1	**1.2**	**2**	**3**	**4**	**5.1**	**5.2**	**6.1**	**6.2**
CF	0.0059	0.0061	0.0008	0.0476	0.0056	0.0008	0.8290	0.1131	0.0382	0.2501
CB	0.0020	0.1200	0.0552	0.0326	0.0803	0.0350	0.1255	0.1341	0.0140	0.2938
NL	0.0000	0.0009	+	+	0.0000	0.0002	+	+	+	+
LEN 10	0.0000	0.0000	0.0000	0.0001	0.0023	0.0004	0.0011	0.0056	+	0.0000
ELO	0.0000	0.0001	0.0000	0.0001	0.0023	0.0004	0.0011	0.0042	+	0.0000

* 1.1—Sipma Z-569/1 Farma II—dried green fodder; 1.2—Sipma Z-569/1 Farma II—spring barley straw; 2—UNIA DF 1,7 Zd—dried green fodder; 3—UNIA DF 1,8 Dd—dried green fodder; 4—Sipma PS 1211 FARMA PLUS—winter triticale straw; 5.1—Warfama Z-543—spring barley straw; 5.2—Warfama Z-543—winter wheat straw; 6.1—Metal-Fach Z-562—winter rye straw; 6.2—Metal-Fach Z-562—winter wheat straw. + the round baler wound the same number of net layers onto the bale, or the length of 10Δ and the elongation were the same for each bale.

The assumption about the distribution normality of all the results for the round balers and the bulk materials collected by them was not met. Therefore, correlations between the bale number and the results of the measurements of respective features were determined using the non-parametric Spearman's rank correlation (Table 3). All determined coefficients have values close to zero, so it can be concluded that there is no relationship. Therefore, the number of bales made by round balers did not affect the quality of net wrapping.

Table 3. Dependence of features on bale number—Spearman's rank correlations.

Variables	CF	CB	NL	LEN 10Δ	ELO
Bale number	0.093653	0.044672	−0.162239	0.156588	0.148954

This was followed by the calculation of the descriptive statistics (Table 3) of all analyzed features: mean, minimum and maximum value, standard deviation and standard error. The calculation results led to the conclusion that the baling presses picked the net correctly, and the set number of layers in most of the tested baling presses and collected bulk materials did not change. This proves that the factory settings of the machines made by the operator before the start of harvesting do not change during operation. The net is correctly picked by the wrapping devices installed in the round balers, regardless of their design. During the wrapping operation, there is no slipping effect of the thinner net (Tama LT) on the feeding rollers of the net bale wrapping device. The number of net wraps of bales for the Sipma Z-569/1 Farna II baler during the harvest of spring barley straw was 2.2. For the UNIA DF 1.7 Zd machine during the harvest of dried green fodder, the number of net wraps of bales was 2.6. For the Warfama Z-543 baler during the harvest of spring barley and winter wheat straw, the number of net layers was 2.5. Very small differences in the number of net layers placed on successive cylindrical bales were found only in three cases: Sipma PS 1211 FARMA PLUS round balers when harvesting winter triticale straw, UNIA Df 1.8 Dd when harvesting dried green fodder and Sipma Z-569/1 Farna II when harvesting dried green fodder. These differences had no effect on the shape of the bales after unloading them from the round baler chamber. The shape of the bales also did not change during the loading into the means of transport, during the transport to the farm, when wrapping the fodder bales with foil or during the storage of winter triticale bales in a heap. Very small differences in the number of net wraps may result from collecting dried fodder, characterized by a higher humidity and generating less tension on the net. Then, when unloading the net-wrapped bales from the round baler chamber, there may be slight movements of the net on the bale. The net loses its ability to move when the threads catch on the stalks of the collected bulk material sticking out on the outer surface of the bale. The number of bale wraps with the net during the harvesting of dried green fodder for the Sipma Z-569/1 Farna II machine ranged from 2.3 to 2.5 (mean 2.4), and for UNIA Df 1.8

Dd, it ranged from 2.7 to 2.8 (mean 2.73). Cylindrical bales formed from dry cereal straw generate more tension on the net than bales of dried green fodder, hay or straw of a higher moisture content. Very small differences in the number of net layers placed on cylindrical bales harvested with the Sipma PS 1211 FARMA PLUS round baler during the harvesting of winter triticale straw resulted from the manual control of the net feeding function in the panel controlling the round baler's parameters. The number of wraps with the net Tama LT on a bale of winter triticale straw ranged from 2.2 to 3.5 (mean 3.64). Furthermore, no changes in the net length or elongation were found in the Metal-Fach Z-562 round baler when harvesting winter rye and winter wheat straw. In this case, the machine was set to the maximum number of net wraps, which was 2.8. The maximum number of bale wraps with the net aimed at the best possible protection of the harvested straw from weather conditions, as it was stored in an outdoor heap.

The smallest and highest mean net coverage of the bale was 112.75 cm and 125 cm, respectively; the differences in coverage between the front and back of the bale were small. The Wilcoxon test was used to compare two dependent populations of the back and front of the bales and demonstrated that these means differ significantly (Tables 4 and 5).

Table 4. Descriptive statistics of round balers collecting dried green fodder.

Trait of Net	Unit	Machine and Bulk Material	No. of Bales	Mean	Minimum	Maximum	SD	SE
CF	cm	Sipma Z-569/1 Farma II dried green fodder	14	121.57	120	123	0.8516	0.2276
CB	cm		14	123.43	122	126	1.1579	0.3095
NL	-		14	2.4	2.3	2.5	0.0555	0.0148
LEN 10Δ	cm		14	61.43	61	62	0.5136	0.1373
ELO	%		14	7.77	7.02	8.77	0.8987	0.2402
CF	cm	UNIA Df 1,8 Dd dried green fodder	12	112.75	112	114	0.6216	0.1794
CB	cm		12	113.08	112	115	0.9962	0.2876
NL	-		12	2.73	2.7	2.8	0.0452	0.0131
LEN 10Δ	cm		12	61.04	59	63	1.5442	0.4458
ELO	%		12	7.09	3.51	10.53	2.7081	0.7818
CF	cm	UNIA DF 1,7 Zd dried green fodder	12	117.67	117	118.5	0.5774	0.1667
CB	cm		12	119.58	117.5	123	2.0542	0.593
NL	-		12	2.6	2.6	2.6	0	0
LEN 10Δ	cm		12	62.17	62	63	0.3257	0.094
ELO	%		12	9.06	8.77	10.53	0.5732	0.1655

The analysis of the results of the length 10Δ on the net before attaching it to the machine hopper and then after wrapping the bale with the net allowed us to determine its percentage elongation. The results of the length 10Δ were the same for the net before it was attached to the machine hopper, which proves the high quality of the net. After wrapping the bale with the net, the length 10Δ leads to the conclusion that the mean values of this parameter are similar for all tested round balers and collected bulk materials. This demonstrates a high correlation between these features. The calculated Spearman's rank correlation coefficient for all observations was $r = 0.9947$. The general correlation coefficient for the tested round balers and collected bulk materials was 0.7807 for variant 5.2. and 1.0000 for the other variants. All correlation coefficients are significant at the significance level of 0.05, so one of the analyzed features, i.e., net elongation, can be selected for further analysis.

The coverage of cylindrical bales with a light type net (Tama LT) on the front and back in the tested variants for the respective machines and bulk materials is presented in the violin plots. Their shape is similar, which may indicate the lack of significant differences between the means of these features in the respective test variants. However, the results of the Wilcoxon test for all observations showed that the means of CF from CB differ significantly. In order to check this assumption, a comparison test was conducted for two means for dependent populations—the Wilcoxon test when the assumption of distribution normality was not met, and the t-distribution test when the assumption was met (Table 6). Since, in some of the analyzed presses and the bulk materials collected by them, the mean CF and CB differ significantly, all these features will be taken into consideration in the following sections.

Table 5. Descriptive statistics of round balers collecting straw after harvesting cereals.

Trait of Net	Unit	Machine and Bulk Material	No. of Bales	Mean	Minimum	Maximum	SD	SE
CF	cm	Sipma Z-569/1 Farma II spring barley straw	30	121.07	120	123	0.9444	0.1724
CB	cm		30	121.83	119	124	1.3667	0.2495
NL	-		30	2.2	2.2	2.2	0	0
LEN 10Δ	cm		30	61.73	61	62	0.3144	0.0574
ELO	%		30	8.3	7.02	8.77	0.5517	0.1007
CF	cm	Warfama Z-543 spring barley straw	12	120.58	116	125	2.4293	0.7013
CB	cm		12	122.17	114	127	3.2427	0.9361
NL	-		12	2.5	2.5	2.5	0	0
LEN 10Δ	cm		12	61.67	61	62	0.4438	0.1281
ELO	%		12	8.19	7.02	8.77	0.777	0.2243
CF	cm	Warfama Z-543 winter wheat straw	12	122.58	120	126	1.6214	0.468
CB	cm		12	123.21	120	126	2.1047	0.6076
NL	-		12	2.5	2.5	2.5	0	0
LEN 10Δ	cm		12	61.38	61	62	0.3108	0.0897
ELO	%		12	7.53	7.02	8.77	0.5836	0.1685
CF	cm	Sipma PS 1211 FARMA PLUS winter triticale straw	19	124.63	122	131	1.921	0.4407
CB	cm		19	125	121	130	2.2361	0.5199
NL	-		19	2.64	2.2	3.5	0.4194	0.0962
LEN 10Δ	cm		19	62.32	62	63	0.342	0.0785
ELO	%		19	9.33	8.77	10.53	0.6019	0.1381
CF	cm	Metal-Fach Z-562 winter rye straw	20	117.38	116	119	0.8717	0.1949
CB	cm		20	117.33	116	119	0.8626	0.1929
NL	-		20	2.8	2.8	2.8	0	0
LEN 10Δ	cm		20	61	61	61	0	0
ELO	%		20	7.02	7.02	7.02	0	0
CF	cm	Metal-Fach Z-562 winter wheat straw	17	116.29	114	120	1.6111	0.3907
CB	cm		17	116.5	114	119	1.4361	0.3483
NL	-		17	2.8	2.8	2.8	0	0
LEN 10Δ	cm		17	60.82	60	61	0.393	0.0953
ELO	%		17	6.71	5.26	7.02	0.6916	0.1677

Table 6. Comparison test of two dependent means between CF and CB; Wilcoxon and Student's *t* depending on the fulfillment of assumptions.

Round Balers—Bulk Material	N	Z/t	*p*
Total	115	4.9885	0.0000
1.1	13	3.1798	0.0015
1.2	26	2.7430	0.0061
2	12	3.0594	0.0022
3	5	1.3484	0.1775
4	16	0.4395	0.6603
5.1	12	T = −2.455	0.0320
5.2	12	T = −0.7718	0.4565
6.1	9	0.00	1.0000
6.2	14	T = −0.7318	0.4749

The dispersion of observations in all plots (Figure 1) varies, which may indicate the heterogeneity of variance. In addition, it is not always symmetrical, which proves that the hypothesis of the distribution compatibility with the normal distribution is rejected (Table 2).

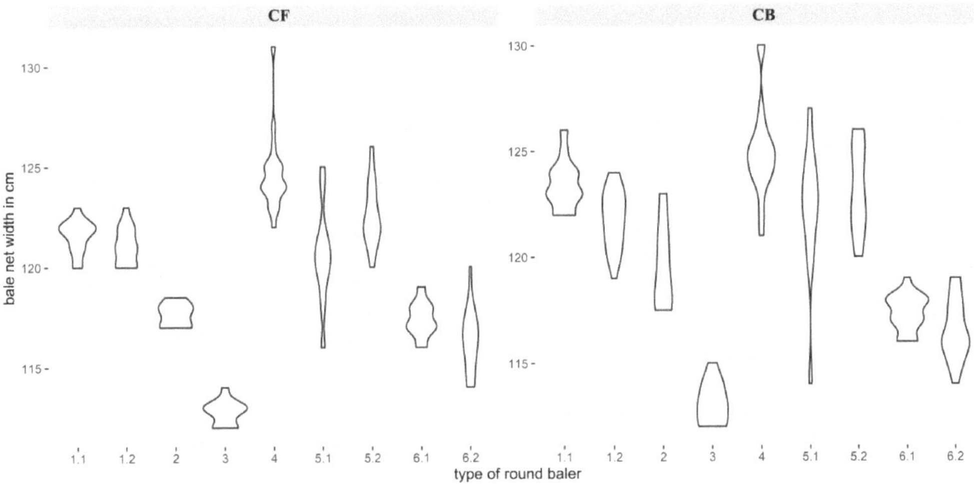

Figure 1. Cover the bales with COVERNET (Tama LT) on the front and back of the bale. 1.1—Sipma Z-569/1 Farma II—dried green fodder; 1.2—Sipma Z-569/1 Farma II—spring barley straw; 2—UNIA DF 1,7 Zd—dried green fodder; 3—UNIA DF 1,8 Dd—dried green fodder; 4—Sipma PS 1211 FARMA PLUS—winter triticale straw; 5.1—Warfama Z-543—spring barley straw; 5.2—Warfama Z-543—winter wheat straw; 6.1—Metal-Fach Z-562—winter rye straw; 6.2—Metal-Fach Z-562—winter wheat straw.

The experimental system was non-orthogonal (Table 1), and, therefore, the comparison was made for levels of one factor that was a combination of two factors: press and bulk material. Since the assumption of the homogeneity of variance for the compared groups was not met for any of the features (p-value < 0.05), the non-parametric Kruskal–Wallis test was used.

The results of the non-parametric analysis of variance are shown in Table 7. No significant differences in terms of all features were found between the balers of the same manufacturer collecting different bulk materials.

There were no significant differences between the forage harvesting machines in terms of the elongation of the net while wrapping the bales. The lowest net elongation of 6.71% was found for the Metal–Fach Z-562 baler when harvesting winter wheat straw. The greatest elongation of the net, amounting to 9.33%, was found for the Sipma PS 1211 FARMA PLUS machine during the harvesting of straw from winter triticale. However, significant differences in the wrapping of the bales with the net in the front and back were found between the Sipma Z-569/1 Farma II baler and the UNIA DF 1.8 Dd baler, which collected dried forage. Wrapping the bales by the UNIA DF 1.8 Dd round baler over a smaller width (CF = 112.75 and CB = 112.08) was directly caused by the design of the wrapping device in the machine. During bale wrapping, the fed net was additionally narrowed on the guiding roller, leading to its distribution over a narrower width. Round balers that collected spring barley straw did not differ significantly in their analyzed features. On the other hand, the balers harvesting winter wheat straw differed in bale wrapping in the front and back and did not differ in the case of the net elongation. In the Metal-Fach Z-562 baling press, a metal roller placed in guides was used to brake the roll and to tension the net. In this case, the acting force was always the same and could not be changed. Overall, out of the 36 pairs of compared factor levels, 19, 20 and 25 pairs did not differ in terms of the front coverage (CF), back coverage (CB) and net elongation (ELO), respectively, which is the majority of the pairs compared.

Table 7. Kruskal–Wallis test *p*-value for multiple (two-sided) comparisons between round baler and bulk material levels.

Coverage Front (CF)	**General Hypothesis Test H (8. N = 148) = 121.1490 p = 0.000**								
	1.1	**1.2**	**2**	**3**	**4**	**5.1**	**5.2**	**6.1**	**6.2**
1.1	-	1.0000	0.1949	0.0000	0.2361	1.0000	1.0000	0.018	0.0016
1.2	1.0000	-	0.0753	0.0000	0.0231	1.0000	1.0000	0.0018	0.0001
2	0.1949	0.0753	-	0.8696	0.0000	1	0.0046	1.0000	1.0000
3	0.0000	0.0000	0.8696	-	0.0000	0.0004	0.0000	1.0000	1.0000
4	0.2361	0.0231	0.0000	0.0000	-	0.05	1.0000	0.0000	0.0000
5.1	1.0000	1.0000	1.0000	0.0004	0.05	-	1.0000	0.2381	0.0329
5.2	1.0000	1.0000	0.0046	0.0000	1.0000	1.0000	-	0.0001	0.0000
6.1	0.0180	0.0018	1.0000	1.0000	0.0000	0.2381	0.0001	-	1.0000
6.2	0.0016	0.0001	1.0000	1.0000	0.0000	0.0329	0.0000	1.0000	-
Coverage back (CB)	**General hypothesis test H (8. N = 148) = 114.7640 p = 0.000**								
	1.1	**1.2**	**2**	**3**	**4**	**5.1**	**5.2**	**6.1**	**6.2**
1.1	-	1.0000	0.1415	0.0000	1.0000	1.0000	1.0000	0.0000	0.0000
1.2	1.0000	-	1.0000	0.0000	0.0664	1.0000	1.0000	0.0019	0.0003
2	0.1415	1.0000	-	0.0628	0.0016	1.0000	0.4442	1.0000	1.0000
3	0.0000	0.0000	0.0628	-	0.0000	0.0000	0.0000	1.0000	1.0000
4	1.0000	0.0664	0.0016	0.0000	-	1.0000	1.0000	0.0000	0.0000
5.1	1.0000	1.0000	1.0000	0.0000	1.0000	-	1.0000	0.0103	0.0022
5.2	1.0000	1.0000	0.4442	0.0000	1.0000	1.0000	-	0.0004	0.0001
6.1	0.0000	0.0019	1.0000	1.0000	0.0000	0.0103	0.0004	-	1.0000
6.2	0.0000	0.0003	1.0000	1.0000	0.0000	0.0022	0.0001	1.0000	-
Elongation (ELO)	**General hypothesis test H (8. N = 148) = 89.82629 p = 0.0000**								
	1.1	**1.2**	**2**	**3**	**4**	**5.1**	**5.2**	**6.1**	**6.2**
1.1	-	1.0000	0.0974	1.0000	0.0029	1.0000	1.0000	1.0000	0.6661
1.2	1.0000	-	1.0000	1.0000	0.1086	1.0000	0.9576	0.0005	0.0002
2	0.0974	1.0000	-	0.4719	1.0000	1.0000	0.0186	0.0000	0
3	1.0000	1.0000	0.4719	-	0.0329	1.0000	1.0000	0.4861	0.2511
4	0.0029	0.1086	1.0000	0.0329	-	0.2774	0.0004	0.0000	0.0000
5.1	1.0000	1.0000	1.0000	1.0000	0.2774	-	1.0000	0.0634	0.0307
5.2	1.0000	0.9576	0.0186	1.0000	0.0004	1.0000	-	1.0000	1.0000
6.1	1.0000	0.0005	0.0000	0.4861	0.0000	0.0634	1.0000	-	1.0000
6.2	0.6661	0.0002	0.0000	0.2511	0.0000	0.0307	1.0000	1.0000	-

$p < 0.05$ means significant differences.

The introduction a new generation of lightweight nets for wrapping bulk materials to the Polish market of agricultural means of production has resulted in a gradual reduction in the consumption of HDPE for the production of agricultural nets in subsequent years (2017–2022) (Table 8). This can also reduce the weight of waste, which is about twice as high because of the plant material, soil and moisture affecting the net.

Table 8. Estimated HDPE consumption data in Poland for the production of agricultural mesh in the years 2017–2022 in Mg.

Year	Consumption of HDPE	HDPE Consumption without LT Technology	Saving HDPE	Reducing the Weight of Waste *
2017	6352	6795	443	885
2018	5589	6039	450	900
2019	5369	5841	473	945
2020	6453	7046	593	1185
2021	6001	6646	645	1290
2022	5565	6270	705	1410
Total	**35,329**	**38,636**	**3308**	**6615**

* 50% of the *waste in the used net is: soil, plant residues from silage, straw or hay and water.

4. Conclusions

No studies of a similar nature were found in the available literature, which would present the results of a new generation of LT agricultural netting for packing bulk materials. There are also no data in the context of the environmental burden caused by agricultural nets. Based on the tests conducted for nine variants of different models of round balers and the different bulk materials collected by them, it can be concluded that, in each case, the net (Tama LT) wrapped the cylindrical bales well or very well. The Tama LT net has received a positive assessment. In every variant tested, the Tama LT net held the compressed material well, maintaining its form shaped in the baler chamber. When the machine was equipped with a net wrapping device for adjusting the net tension during the wrapping, the bales were completely covered with the net. By increasing the braking force of the Tama LT net roll, it was possible to wrap the outer surface around the bale circumference, as well as the side surfaces, which is very beneficial. In this case, the edges of the bales are even, so they can be wrapped more precisely when harvesting green fodder for silage. Then, better conditions are provided for the silage of green fodder to obtain a better quality of the fodder. The mean elongation of the COVERNET during the bale wrapping was over 8% for the tested machines and harvested materials. The Tama LT net was efficiently picked and fed by the wrapping devices of various designs used in the tested round balers. The Tama LT net also ensured the trouble-free operation of the balers. The tests confirmed the usefulness of the new generation of agricultural nets (Tama LT) for wrapping various agricultural bulk materials of various humidities. There were no cases of bale deformation or net breaking when the bales were unloaded from the bale chamber onto the field surface. There was also no net breaking and no bale deformation during the bale loading, transport and storage in heaps. Therefore, the net manufactured in the LT technology with a breaking strength of 2.5 kN guarantees that it can be used for wrapping bales of various agricultural bulk materials. The new-generation agricultural netting can be used for baling presses equipped with wrapping devices. The net produced in light technology (Tama LT) is therefore a high-quality product that is usable under various harvesting conditions, despite the lower weight per 1 m of the net.

There is currently no developed technology for recycling used agricultural netting after its use in the harvesting of agricultural bulk materials. The introduced new technology for the agricultural net production allows for reducing CO_2 emissions by 25% compared to the standard technology. This is directly related to the lower demand for granulate at the stage of agricultural net production in modern LT technology. At the same time, there is an urgent need to develop and implement in practice a technology for recovering used agricultural nets and converting them into granules that can be used in their production.

Author Contributions: Conceptualization, I.K.; Methodology, I.K.; Validation, B.Z. and P.R.; Formal analysis, K.K.; Data curation, I.K. and B.Z.; Writing—original draft, I.K.; Writing—review & editing, P.R. All authors have read and agreed to the published version of the manuscript.

Funding: This research received no external funding.

Institutional Review Board Statement: The study did not require ethical approval.

Data Availability Statement: Data supporting the reported results can be found in the Department of Agronomy at the University of Life Sciences.

Conflicts of Interest: The authors declare no conflict of interest.

References

1. Dőring, T.F.; Kovacs, G.; Wolfe, M.S.; Murphy, K. Evolutionary plant breeding in cereals—Into new era. *Sustainability* **2011**, *3*, 1944–1971. [CrossRef]
2. Gacek, E.S. Modyfikacje prac hodowlanych i doświadczalnictwa odmianowego dla potrzeb zrównoważonych, niskonakładowych i ekologicznych systemów gospodarowania w rolnictwie. *Biul. Inst. Hod. I Aklim. Roślin* **2017**, *282*, 139–150. (In Polish)
3. Wicki, L. Postęp w plonowaniu odmian pszenicy ozimej i żyta w doświadczeniach odmianowych w Polsce. *Rocz. Nauk. Stowarzyszenia Ekon. Rol. I Agrobiz.* **2017**, *XIX*, 224–230. (In Polish)
4. Arseniuk, E.; Oleksiak, T. Postęp w hodowli głównych roślin uprawnych w Polsce i możliwości jego wykorzystania do 2020 roku. *Stud. I Rap. IUNG-PIB* **2009**, *14*, 293–306. (In Polish) [CrossRef]
5. Mańkowski, D.R. Postęp biologiczny w hodowli, nasiennictwie i produkcji ziemniaka w Polsce. Część I. Przegląd ilościowych metod oceny postępu hodowlanego i odmianowego. *Biul. Inst. Hod. I Aklim. Roślin* **2009**, *251*, 153–173. (In Polish)
6. Trethowan, R.M.; van Ginkel, M.; Rajaram, S. Progress in breeding wheat for yield and adaptation in global drought affected environments. *Crop Sci.* **2002**, *42*, 1441–1446. [CrossRef]
7. Ustun, A.; Allen, F.L.; English, B.C. Genetic progress in soybean of the U.S. Midsouth. *Crop Sci.* **2001**, *41*, 993–998. [CrossRef]
8. Święcicki, W.K.; Surma, M.; Koziara, W.; Skrzypczak, G.; Szukała, J.; Bartkowiak-Broda, I.; Zimny, J.; Banaszak, Z.; Marciniak, K. Nowoczesne technologie w produkcji roślinnej—Przyjazne dla człowieka i środowiska. *Pol. J. Agron.* **2011**, *7*, 102–112.
9. Mikołajczyk, K.; Dabert, M.; Nowakowska, J.; Podkowinski, J.; Poplawska, W.; Bartkowiak-Broda, I. Conversion of the RAPD OPC021150 marker of the Rfo restorer gene into a SCAR marker for rapid selection of oilseed rape. *Plant Breed.* **2008**, *127*, 647–649. [CrossRef]
10. Dawson, J.C.; Rivière, P.; Berthellot, J.F.; Mercier, F.; de Kochko, P.; Galic, N.; Pin, S.; Serpolay, E.; Thomas, M.; Giuliano, S.; et al. Collaborative plant breeding for organic agricultural systems in developed countries. *Sustainability* **2011**, *3*, 1206–1223. [CrossRef]
11. Murphy, K.M.; Lammer, D.; Lyon, S.R.; Carter, B.; Jones, S.S. Breeding for organic and low-input farming systems: An evolutionary-participatory breeding method for inbred cereal grains. *Renew. Agric. Food Syst.* **2005**, *20*, 48–55. [CrossRef]
12. Finckh, M.R. Integration of breeding and technology into diversification strategies for disease control in modern agriculture. *Eur. J. Plant Pathol.* **2008**, *121*, 399–409. [CrossRef]
13. Kotecki, A. Climate change for the European Green Deal. In Proceedings of the Scientific Conference: The Role of Agricultural Sciences in the Implementation of the Concept of a Sustainable Food System "from Farm to Fork", Chełm, Poland, 7–8 June 2022. (In Polish).
14. Hammond, G.; Jones, C.; Lowrie, F.; Tse, P. *The Inventory of Carbon and Energy (ICE)*; University of Bath: Bath, UK, 2011; p. 128, © BSRIA BG 10/2011; ISBN 978 0 86022 703 8.
15. Briassoulis, D.; Mistriotis, A.; Eleftherakis, D. Mechanical behaviour and properties of agricultural nets Part I: Testing methods for agricultural nets. *Polym. Test.* **2007**, *26*, 822–832. [CrossRef]
16. Briassoulis, D.; Mistriotis, A.; Eleftherakis, D. Mechanical behaviour and properties of agricultural nets. Part II: Analysis of the performance of the main categories of agricultural nets. *Polym. Test.* **2007**, *26*, 970–984. [CrossRef]
17. Castellano, S.; Scarascia Mugnozza, G.; Russo, G.; Briassoulis, D.; Mistriotis, A.; Hemming, S.; Waaijenberg, D. Plastic nets in agriculture: A general review of types and applications. *Appl. Eng. Agric.* **2008**, *24*, 799–808. [CrossRef]
18. Chodak, I. High modulus polyethylene fibres: Preparation, properties and modification by crosslinking. *Prog. Polym. Sci.* **1998**, *23*, 1409–1442. [CrossRef]
19. Ward, I.M. Recent developments in the science and technology of high modulus flexible polymers. *Macromol. Symp.* **1995**, *100*, 1–14. [CrossRef]
20. Barham, P.J.; Keller, A. Review: High-strength polyethylene fibres from solution and gel spinning. *J. Mater. Sci.* **1985**, *20*, 2281-2032. [CrossRef]
21. Tayyab, H.; Ibnelwaleed, A.H. Effect of short chain branching of LDPE on its miscibility with linear HDPE. *Macromol. Mater. Eng.* **2004**, *289*, 198–203. [CrossRef]
22. Tama Polska Sp. z o.o. *CPA Technical*; Crop Packaging Association: Alton, UK, 2011; Volume 1, p. 4. (In Polish)

23. Tama Polska Sp. z o.o. *CPA Technical*; Crop Packaging Association: Alton, UK, 2011; Volume 2, p. 4. (In Polish)
24. Tama Polska Sp. z o.o. *CPA Technical*; Crop Packaging Association: Alton, UK, 2012; Volume 3, p. 4. (In Polish)

agriculture

MDPI

Article

Selecting the 'Sustainable' Cow Using a Customized Breeding Index: Case Study on a Commercial UK Dairy Herd

Matt J. Bell * and Greta-Marie Jauernik

Animal and Agriculture Department, Hartpury University, Gloucester GL19 3BE, UK
* Correspondence: matt.bell@hartpury.ac.uk

Abstract: The aim of the current study was to investigate using a customized profit and carbon total merit index to identify sustainable milking cows and herd replacements within a commercial dairy herd. Balancing the economic, social and environmental aspects of milk production has gained interest given the increasing global demand for milk products. Furthermore, a farm-level customized breeding index with farm-derived weightings for biological traits would incorporate the effect of the farm environment. This study used a Markov chain approach to model a commercial dairy herd in the UK between the years 2017 and 2022. Production, financial, genetic and nutritional data for the herd were used as input data. The model derived the economic (GBP per unit) and carbon values (kilograms CO_2-eq. emissions per unit) for a single phenotypic increase in milk volume, milk fat yield, milk protein yield, somatic cell count, calving interval and lifespan, which were used in a profit and carbon index. The study proposed a methodology for selecting individual milking cows and herd replacements based on their potential to increase herd profitability and reduce carbon emissions as a means to identify more sustainable animals for a given farm environment. Of the 370 cows and herd replacements studied, 76% were classified as sustainable with a desirable increase in profit and reduction in carbon emissions. Customized breeding indices with trait weightings derived from the farm environment and selecting individual animals on economic and carbon metrics will bring permanent and cumulative improvements to the sustainability of milk production with appropriate nutrition and management. The approach used can be applied to any commercial farm to select animals that are more sustainable.

Keywords: dairy systems; biological traits; profit; greenhouse gas emissions

1. Introduction

Global milk production and the number of milking cows are expected to keep increasing with the continued demand for dairy products [1]. About two-thirds of the carbon footprint of fresh milk is associated with the animal in the form of enteric and manure methane [1]. Therefore, the mitigation of greenhouse gas emissions is a priority if we are to achieve net zero targets in the future. In the UK, the emissions per unit of milk produced by dairy cows has been reducing by about 1% per annum over the last few decades with improved efficiencies, primarily due to better genetic selection and nutrition. However, the emissions per cow are estimated to increase by 1.0% per annum due to increased production [2]. The production efficiencies per unit product have been achieved by increasing productivity and gross efficiency (i.e., the ratio of yield of milk to resource input) with a dilution in the maintenance cost of animals in the system and decreasing the number of animals needed to produce the same amount of product [3,4]. Previous studies have highlighted that sustainable livestock breeding should aim to increase productivity while reducing negative environmental effects and improving livestock welfare [5,6]. Furthermore, Richardson et al. [7] identified that the use of genetic selection indexes should be explored further to reduce carbon emissions from livestock.

Given that genetic selection is permanent and cumulative with time, it is recognized as a cost-effective option for mitigating greenhouse gas emissions from livestock production [3]. However, the balance between productivity and resource efficiency has been a challenge for decades [8–10]. Genetic selection in livestock, such as dairy cows, has increasingly relied on a balance between the selection of more heritable production traits (e.g., kilograms milk, kilograms fat and protein) given their market value and fewer heritable fitness traits (e.g., lameness, mastitis, fertility and lifespan), which have welfare implications. The selection of dairy cows based on health, fertility and overall survival has been found to bring profitable reductions in the carbon footprint of milk with improved resource efficiency [2,11], which are all important social aspects for consumer confidence in livestock farming. Therefore, with more emphasis on fitness rather than production traits, the health and fertility of dairy cows is expected to improve, along with the carbon footprint of milk production. For this very reason, several countries (France, Italy, Germany, Switzerland, Belgium, Australia, the United States, the UK, Nordic countries, Ireland and The Netherlands) put more emphasis on fitness traits (>50% weighting) than milk production traits (milk, fat and protein yield) in their national breeding programmes [12]. In addition to these changes, breeding programmes are not only focused on economics but also health and environmental objectives. The use of a customized selection index, where a producer creates economic or other weights tailored to the farm environment, rather than the use of a nationally-derived breeding index and weightings (such as the economic Profitable Lifetime Index for dairy cows in the UK), may be more appropriate for traits associated with health and environmental goals [10].

The objective of the current study was to investigate the use of a customized profit and carbon total merit index to identify sustainable milking cows and herd replacements within a commercial dairy herd.

2. Materials and Methods

2.1. Data

Average production records between the years 2017 and 2022 were obtained for the Hartpury University Dairy herd, which is a commercially run and milk-recorded herd in the UK (Table 1). The herd consisted predominantly of autumn calving Holstein Friesian cattle with 60 one-year-old heifers, 72 first-lactation heifers and 238 older milking cows (up to eight lactations). The production (Table 1), financial, genetic and nutritional data for the herd and herd replacements were obtained.

Table 1. Average production values per lactation.

Trait	Units	Value
Milk volume	kg	8909
Milk fat yield	kg	347
Milk protein yield	kg	285
Lifespan	lactations	2.2
Somatic cell count	'000 cells/mL	129
Calving interval	days	368
Enteric CH_4 [1]	kg	146
Manure CH_4	kg	55
Manure N_2O [2]	kg	7
CO_2 equivalent emissions	tonnes	7.3

[1] Enteric CH_4 emissions per kg dry matter (DM) intake were estimated by: CH_4 (g/kg DM intake) = 0.046 × DOMD − 0.113 × ether extract (both g/kg DM) − 2.47 × (feeding level − 1), where DOMD is digestible organic matter in the dry matter and feeding level is metabolizable energy intake as multiples of maintenance energy requirements. [2] Direct and indirect N_2O emissions from stored manure and application of faeces, urine and manure and land applications of manure (from leaching and atmospheric deposition of nitrogen from NOx and NH_3) as used by the National GHG Inventory for agriculture in the UK.

The herd income and variable costs (Table 2) were used to derive the gross profit or loss per cow in a partial budget.

Table 2. Income and output costs (GBP) calculated for the herd per cow.

	Value
Income	GBP
Milk sales [1]	3029.24
Calves [2]	66.97
Culls [3]	141.70
Less	
Replacements [4]	877.70
Total Output	2013.43
Variable costs	
Feed	1298.51
Dairy supplies [5]	412.93
Health problems	183.01
Fertility	26.48
Total variable costs	1920.93
Gross Margin	439.28

[1] The average milk price was 34 p/L. [2] Average calf value of GBP 2.50 per kilogram body weight. [3] Average cull cow value of GBP 0.50 per kilogram body weight. [4] Average herd replacement cost of GBP 2.20 per kilogram body weight. [5] Average cost of GBP 0.05 per litre milk for recording, parlour consumables and sundries.

The predicted transmitting abilities (PTA) for each animal were calculated using the most recent genetic evaluations from August 2022 (Table 3). The PTA represents a prediction of the increased or reduced unit change in a trait that a cow will transmit to their progeny relative to the national average PTA of zero for the same month. The herd represented a wide range of positive and negative PTAs for production (milk volume, milk fat and milk protein) and fitness (somatic cell count, lifespan and fertility) traits included in this study.

Table 3. Average (s.d.) predicted transmitting ability per animal.

Trait	Units	Average	Min	Max
Milk volume	kg	74 (254)	−619	684
Milk fat	kg	2.7 (11)	−27	30
Milk protein	kg	3.2 (7.9)	−20	20
Somatic cell count	%	−1.8 (6.6)	−20	15
Lifespan	days	52 (36)	−92	122
Fertility	days	3.2 (3.9)	−11	12

The diet for a herd replacement heifer and lactating cow contained pasture, grass silage and concentrate feed (Table 4).

Table 4. Content and composition of the diet of a heifer replacement and lactating cow.

Nutrient Content	Units	Replacement	Lactating Cow
Crude protein (CP)	g/kg DM	142	203
Neutral detergent fibre (NDF)	g/kg DM	483	367
Ether extract	g/kg DM	49	42
Ash	g/kg DM	78	67
Metabolisable energy (ME)	MJ/kg DM	10.6	12.2
Feeding level [1]		2.5	4.7
Digestible organic matter in dry matter (DOMD) [1]	g/kg DM	661	711
Organic matter digestibility (OMD) [1]	% of OM	71.7	76.2
Digestible CP [1]	g/kg DM	85	143
Methane [1]	g/kg DM	21.1	18.8

Table 4. *Cont.*

Nutrient Content	Units	Replacement	Lactating Cow
Composition			
Pasture	%	33	33
Conserved forage	%	50	32
Concentrate	%	17	35

[1] The DOMD was estimated from Wainman (1981) as: DOMD (g/kg DM) = 472.49 × ln(ME) − 437.69; % OMD = [DOMD/(1000 − ash)] × 100; Digestible CP (g/kg DM) was estimated by the rearranged equation of Wang et al. (2009) as = CP − [((ln((OMD/100 − 0.899)/−0.644) × 100)/−0.5774)/1000] × ((1000 − ash) − DOMD); Enteric CH_4 emissions were estimated as: CH_4 (g/kg DM intake) = 0.046 × DOMD − 0.113 × ether extract − 2.47 × (feeding level − 1) with the feeding level being estimated from metabolizable energy intake as multiples of maintenance energy required.

2.2. Modelled Current and Adjusted Herd

The production, financial and nutritional data for the herd were used as inputs for an existing bioeconomic model that describes the nutrient partitioning of a cow over its productive life using a Gompertz growth curve (growth rate of 0.0033 kg protein/day). For more detail, see Bell et al. [2,11].

The Markov chain stochastic framework describes the herd structure as 11 age groups including life prior to entering the herd and from lactations one to 10 to cover the likely lifespan of a milking cow. This approach allows for a change in lifespan and herd structure to be assessed. The herd is described as a vector of states (s) that cows occupy at a given point in time [13], and each age group was included in the current study. Briefly, the vector of states at time t is multiplied by a matrix of transition probabilities (s × s) to generate a vector of states at time t + 1. The probability of a cow surviving to the next lactation (from lactation n to n + 1 and from lactation 1 to n) was dependent on survival during the current lactation. The model allows herd level data to be combined and cow biological traits to be adjusted in order to test the effect of adjusting the animal traits of interest on the key production, environmental and economic metrics as described below. Replacements joined the herd at 741 days of age on average, and sexed semen was used to breed herd replacements.

2.3. Feed Intake and Nutritional Requirements

The energy requirements (of herd replacements and lactating cows) for maintenance, growth, pregnancy, activity and lactation (E_{total}) are assumed to be achieved, and feed intake is always sufficient to achieve energy requirements. The metabolizable energy (ME, MJ/d) required for maintenance (E_{maint}), gain or loss of body protein (E_p) and lipid (E_l), pregnancy (E_{preg}), activity (E_{act}) and lactation (E_{lact}) for the average cow in the herd presented in Table 5 were based on average production data (Table 1).

Table 5. Percentage of total metabolizable energy (% of ME) for a heifer replacement and the average lactating dairy cow for maintenance (E_{maint}), protein (E_p) and lipid growth (E_l), pregnancy (E_{preg}), activity (E_{act}) and milk production (E_{lact}) for the current herd situation.

Energy Requirement	Replacement	Lactating Cow
E_{maint}	50.9	25.7
E_p	15.3	0.1
E_l	24.6	0.3
E_{preg}	4.0	2.4
E_{act}	5.1	2.6
E_{lact}	0.0	68.9
Total (E_{total} MJ)	38,890	76,556

The average total ME (E_{total}) requirement for each age group was used to calculate the feed intake (kg DM) of an animal (Equation (1)):

$$\text{Feed intake (kg DM)} = E_{total} \times 1/(ME - 0.616 \times ECH_4 - 3.8/FE - 29.2 \times DCP/6.25) \quad (1)$$

where ME, ECH_4 and FE are the metabolizable, enteric CH_4 and faecal energy (all MJ kg^{-1} DM), respectively, and DCP is the digestible crude protein (kg/kg DM). The values of 0.616, 3.8 and 29.2 are the heat increments associated with fermentation, faeces and DCP.

The total DM intake multiplied by ME content (Table 4) and cost per unit ME of the diet allowed the cost of feed consumed by each age group to be estimated, with pasture costing GBP 0.003 per MJ ME, grass silage costing GBP 0.009 per MJ ME and concentrates costing GBP 0.026 per MJ ME).

2.4. *Changes in Profit and Carbon Emissions*

The main greenhouse gases included were enteric and manure CH_4 and direct and indirect N_2O emissions from stored manure and application of faeces, urine and manure and land applications of manure (from leaching and atmospheric deposition of nitrogen from NOx and NH_3) as used by the National GHG Inventory for agriculture in the UK [14]. The IPCC [15] Tier II methodology was used to estimate manure CH_4 and N_2O emissions (from N excretion) from storage, as well as manure on fields. The estimated amount of N excreted by the animal was modelled to partition into faeces (N intake − digested N intake) and urine (N intake − (N retained + N in faeces)). Undigested organic matter in the diet (1 − digestible organic matter kg/kg) was used to estimate the other volatile solids in the manure. Emissions were expressed as CO_2-eq. emissions per cow. Kilograms of CO_2-eq. emissions for a 100-year time horizon were calculated using conversion factors from CH_4 to CO_2 of 25 and from N_2O to CO_2 of 298 [15]. The loss of dietary energy as enteric CH_4 was calculated using Equation (2) by Bell et al. [16]:

$$\text{CH4 (g/kg DM intake)} = 0.046 \times \text{DOMD} - 0.113 \times \text{ether extract (both g/kg DM)} - 2.47 \times (\text{feeding level} - 1) \quad (2)$$

where DOMD is the digestible organic matter in the dry matter and the feeding level is the metabolizable energy intake as multiples of the maintenance energy requirements.

The economic value and emission intensity in kilograms of CO_2-eq. emissions per cow were calculated by a single unit increase in the following biological traits of interest: milk volume, fat yield, protein yield, somatic cell count, calving interval (fertility) and lifespan. Responses to changes are quantified by calculating differences between the current herd situation and an adjusted situation due to a single unit change in each trait.

3. Results and Discussion

The herd studied represented a typical UK dairy herd with summer grazing and supplementary feeding (conserved forage and concentrate) and winter housing on solely conserved forage and concentrate feed (Table 4). The milk volume (8909 kg), milk fat yield (347 kg) and milk protein yield (285 kg) per cow (Table 1) were similar to the production of the UK average herd (8965 kg, 358 kg and 290 kg, respectively), but the average age of cows was lower at 2.2 lactations compared to 2.9 for the average UK herd [2,17]. The average calving interval of 368 days reflected the seasonal calving pattern of the herd studied. In terms of the genetic background of the cows in the case study herd, the data represent the production and fitness traits with a similar magnitude of negative and positive values for PTAs. The traits included are commonly available from herd genetic evaluations and have importance when applying economic and carbon weightings to a total merit index [2]. When genomic predictions for efficiency traits such as feed intake and methane output become routinely available for dairy genetic evaluations, then these traits can be included given their importance with regard to herd profit and carbon emissions [18,19]. After a single unit increase in each trait for the modelled herd, the following economic (GBP/cow;

Equation (3)) and carbon (CO_2-eq./cow; Equation (4)) weightings were derived to obtain a profit and carbon index based on the following trait PTAs:

$$\text{Profit index (GBP per cow)} = -0.08 \times \text{milk volume PTA} + 3.24 \times \text{milk fat yield PTA} + 3.91 \times \text{milk protein yield PTA} - 1.50 \times \text{SCC PTA} + 3.61 \times \text{calving interval PTA} + 1.48 \times \text{lifespan PTA} \quad (3)$$

$$\text{Carbon index (GBP per cow)} = 0.12 \times \text{milk volume PTA} + 5.55 \times \text{milk fat yield PTA} + 1.04 \times \text{milk protein yield PTA} + 0.84 \times \text{SCC PTA} - 16.83 \times \text{calving interval PTA} - 3.65 \times \text{lifespan PTA} \quad (4)$$

After applying both the profit and carbon indices to the PTAs for the 370 cows and heifers in the herd studied, 76% of the animals were classified as sustainable with a positive profit and negative carbon indices values based on their genetic background (i.e., animals in the top left corner of Figure 1). Overall, the average profit index was GBP 102/cow and the average carbon index was −195 kg CO_2-eq./cow. The values derived using the profit index (Equation (3)) were highly correlated (r = 0.85) with the Profitable Lifetime Index values calculated viaq UK national genetic evaluations for the same cows in the current study. This similarity is largely due to both profit indices having 50% weighting on production traits and 50% on fitness traits, even though the Profitable Lifetime Index includes more biological traits. Traditional breeding indices aim to identify suitable females or sires for breeding replacements. Kelleher et al. [20] proposed a lifetime profit index (cow own worth, COW) to help identify the most profitable dairy cows in a herd to aid replacement management rather than engaging in selection based on their breeding potential. The COW index was based on the expected economic performance in current and future lactations with the total genetic merit (i.e., additive and nonadditive genetic merit) of the animal as well as both permanent and temporary (e.g., season of calving, parity) environmental effects. Higher ranking cows on the COW index were associated with more milk and milk solids and calved earlier than lower ranking cows. Dunne et al. [21] proposed a beef breeding index framework based on the future profit potential of female beef cattle with the aim of identifying animals for culling. The approach was a modified version of the index of Kelleher et al. [20] and included genetic and non-genetic effects associated with each female. van de Heide et al. [22] also proposed an approach using genomic breeding values and phenotypic information to predict dairy cattle survival. The authors found that combining genetic and phenotypic information resulted in better predictions of survival. The indices developed in the current study were also based on genetic and phenotypic cow performance over a lifetime and include the period prior to entering the milk herd. The selection of more sustainable herd replacements with both profit and carbon indices (Equations (3) and (4)) will enhance the cow health, fertility and lifespan in the herd, especially with 25% weighting on production and 75% on fitness traits in the carbon index. Ultimately, poor health and fertility impacts the lifespan of animals, which has great importance for the economic, environmental and social aspects associated with the way livestock systems are managed. The traits included in the profit and carbon indices in the current study were similar to the traits found to be independent of carbon emissions in milk volume, fat yield, protein yield, survival and feed saved [23].

Even though the study was conducted on a single herd, the animals were all managed within the same production system with detailed herd information. Notably, 77% of first lactation heifers, 70% of older milking cows and 100% of the one-year-old replacement heifers were classified as sustainable for the farm studied. As individual animal genetic evaluations and economic plus carbon emissions change with time, the rankings can be updated in time for breeding management. The animals identified as sustainable for the production system in the current study can be prioritised for future herd replacements when planning breeding. This information and the use of customised economic and carbon weightings provides a more targeted selection tool than a national breeding index. This was also noted by Kelleher et al. [20]. The more targeted and tailored selection of livestock on economic and resource efficiency metrics at the individual animal level should improve overall herd performance and the sustainability of milk production. Deriving emission intensity weightings for biological traits is becoming more common for national environmental index use [7,18] rather than solely focusing on economic values.

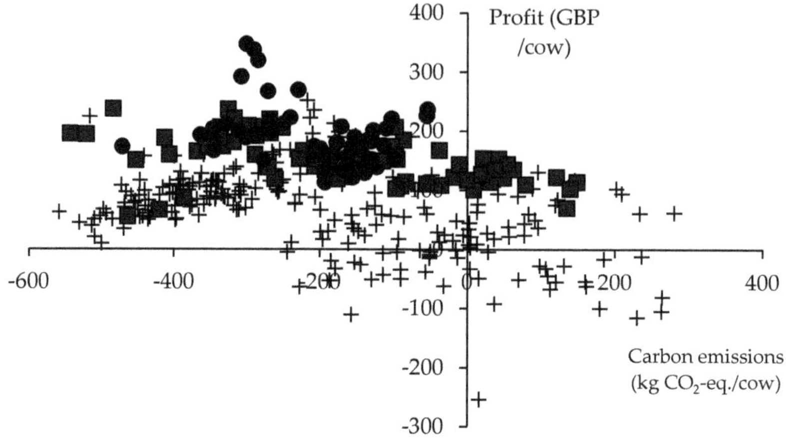

Figure 1. Profit (GBP/cow) and carbon (kg CO_2-eq./cow) index values calculated for milking cows (+), first lactation heifers (■) and one-year-old heifers (•) within the herd studied.

Since changes due to genetic selection are permanent and cumulative with time, it is considered a cost-effective strategy for future profitable reductions in carbon emissions per cow. However, for cows and heifers to achieve their genetic and sustainability potential, they would still require appropriate nutrition and management to meet their requirements. The current study applied a methodology that could be used at the farm level to select female and male animals that are suited to the farm environment, ultimately improving the production efficiency and environmental footprint. The more targeted selection of individual animals will be needed for livestock production systems to be more sustainable in the future. This work should be applied to more farms and national dairy cow populations to help farmers identify animals that are more sustainable.

4. Conclusions

In the present study, 76% of milking cows and heifer replacements studied were classified as sustainable, which was classified as an animal with a desirable positive profit and negative carbon indices values. Customized breeding indices with trait weightings derived for the farm environment and selecting individual animals based on economic and carbon metrics will bring permanent and cumulative improvements to the sustainability of milk production with appropriate nutrition and management. The approach used can be applied to any commercial farm to select animals that are more sustainable.

Author Contributions: Conceptualization, M.J.B.; methodology, M.J.B.; software, M.J.B.; formal analysis, M.J.B.; investigation, M.J.B.; resources, M.J.B.; data curation, M.J.B.; writing—original draft preparation, M.J.B. and G.-M.J.; writing—review and editing, M.J.B. and G.-M.J. All authors have read and agreed to the published version of the manuscript.

Funding: This research received no external funding.

Institutional Review Board Statement: The study did not require ethical approval as it used existing milk recording data.

Data Availability Statement: The datasets analysed are available from the corresponding author upon request.

Acknowledgments: We are grateful for the data obtained from the Hartpury University Dairy Farm.

Conflicts of Interest: The authors declare no conflict of interest.

References

1. FAO; GDP. *Climate Change and the Global Dairy Cattle Sector—The Role of the Dairy Sector in a Low-Carbon Future*; The Food and Agriculture Organization: Rome, Italy, 2018.
2. Bell, M.J.; Garnsworthy, P.C.; Stott, A.W.; Pryce, J.E. Effects of changing cow production and fitness traits on profit and greenhouse gas emissions of UK dairy systems. *J. Agric. Sci.* **2015**, *153*, 138–151. [CrossRef]
3. Bell, M.; Wall, E.; Russell, G.; Simm, G.; Stott, A. The effect of improving cow productivity, fertility, and longevity on the global warming potential of dairy systems. *J. Dairy Sci.* **2011**, *94*, 3662–3678. [CrossRef] [PubMed]
4. Capper, J.L.; Cady, R.A.; Bauman, D.E. The environmental impact of dairy production: 1944 compared with 2007. *J. Anim. Sci.* **2009**, *87*, 2160–2167. [CrossRef] [PubMed]
5. Clay, N.; Garnett, T.; Lorimer, J. Dairy intensification: Drivers, impacts and alternatives. *Ambio* **2020**, *49*, 35–48. [CrossRef] [PubMed]
6. Brito, L.F.; Bedere, N.; Douhard, F.; Oliveira, H.R.; Arnal, M.; Peñagaricano, F.; Schinckel, A.P.; Baes, C.F.; Miglior, F. Review: Genetic selection of high-yielding dairy cattle toward sustainable farming systems in a rapidly changing world. *Animal* **2021**, *15*, 100292. [CrossRef] [PubMed]
7. Richardson, C.M.; Amer, P.R.; Hely, F.; van den Berg, I.; Pryce, J.E. Estimating methane coefficients to predict the environmental impact of traits in the Australian dairy breeding program. *J. Dairy Sci.* **2021**, *104*, 10979–10990. [CrossRef] [PubMed]
8. Rauw, W.M. Immune response from a resource allocation perspective. *Front. Genet.* **2012**, *3*, 267. [CrossRef] [PubMed]
9. del Prado, A.; Mas, K.; Pardo, G.; Gallejones, P. Modelling the interactions between C and N farm balances and GHG emissions from confinement dairy farms in northern Spain. *Sci. Total Environ.* **2013**, *465*, 156–165. [CrossRef] [PubMed]
10. Cole, J.B.; VanRaden, P.M. Symposium review: Possibilities in an age of genomics: The future of selection indices. *J. Dairy Sci.* **2018**, *101*, 3686–3701. [CrossRef] [PubMed]
11. Bell, M.J.; Eckard, R.J.; Haile-Mariam, M.; Pryce, J.E. The effect of changing cow production and fitness traits on net income and greenhouse gas emissions from Australian Dairy systems. *J. Dairy Sci.* **2013**, *96*, 7918–7931. [CrossRef] [PubMed]
12. Eggar-Danner, C.; Cole, J.B.; Pryce, J.E.; Gengler, N.; Heringstad, B.; Bradley, A.; Stock, K.F. Invited review: Overview of new traits and phenotyping strategies in dairy cattle with a focus on functional traits. *Animal* **2015**, *9*, 191–207. [CrossRef] [PubMed]
13. Stott, A.W.; Veerkamp, R.F.; Wassell, T.R. The economics of fertility in the dairy herd. *Anim. Sci.* **1999**, *68*, 49–57. [CrossRef]
14. *UK Greenhouse Gas Inventory (UKGGI). 1990 to 2010 Annual Report for Submission under the Framework Convention on Climate Change*; Defra: London, UK, 2010.
15. Intergovernmental Panel on Climate Change (IPCC). Changes in atmospheric constituents and in radiative forcing. In *Climate Change 2007: The Physical Science Basis. Contribution of Working Group I to the Fourth Assessment Report of the Intergovernmental Panel on Climate Change*; Solomon, S., Qin, D., Manning, M., Chen, Z., Marquis, M., Averyt, K.B., Tignor, M., Miller, H.L., Eds.; Cambridge University Press: Cambridge, UK; New York, NY, USA, 2007; Chapter 2.
16. Bell, M.J.; Eckard, R.; Moate, P.J.; Yan, T. Modelling the effect of diet composition on enteric methane emissions across sheep, beef cattle and dairy cows. *Animals* **2016**, *6*, 54. [CrossRef] [PubMed]
17. Bell, M.J.; Wilson, P. Estimated differences in economic and environmental performance of forage-based dairy herds across the UK. *Food Energy Secur.* **2018**, *7*, e00127. [CrossRef]
18. Amer, P.R.; Hely, F.S.; Quinton, C.D.; Cromie, A.R. A methodology framework for weighting genetic traits that impact greenhouse gas emissions intensity into selection indexes. *Animal* **2018**, *12*, 5–11. [CrossRef] [PubMed]
19. González-Recio, O.; López-Paredes, J.; Ouatahar, L.; Charfeddine, N.; Ugarte, E.; Alenda, R.; Jiménez-Montero, J.A. Mitigation of greenhouse gases in dairy cattle via genetic selection: 2. Incorporating methane emissions into the breeding goal. *J. Dairy Sci.* **2020**, *103*, 7210–7221. [CrossRef] [PubMed]
20. Kelleher, M.M.; Amer, P.R.; Shalloo, L.; Evans, R.D.; Byrne, T.J.; Buckley, F.; Berry, D.P. Development of an index to rank dairy females on expected lifetime profit. *J. Dairy Sci.* **2015**, *98*, 4225–4239. [CrossRef] [PubMed]
21. Dunne, F.L.; Berry, D.P.; Kelleher, M.M.; Evans, R.D.; Walsh, S.W.; Amer, P.R. An index framework founded on the future profit potential of female beef cattle to aid the identification of candidates for culling. *J. Anim. Sci.* **2020**, *98*, skaa334. [CrossRef] [PubMed]
22. van der Heide, E.M.M.; Veerkamp, R.F.; van Pelt, M.L.; Kamphuis, C.; Ducro, B.J. Predicting survival in dairy cattle by combining genomic breeding values and phenotypic information. *J. Dairy Sci.* **2020**, *103*, 556–571. [CrossRef] [PubMed]
23. Richardson, C.M.; Sunduimijid, B.; Amer, P.; van den Berg, I.; Pryce, J.E. A method for implementing methane breeding values in Australian dairy cattle. *Anim. Prod. Sci.* **2021**, *61*, 1781–1787. [CrossRef]

agriculture

MDPI

Article

Organic Farming as a Driver of Environmental Benefits or the Other Way Around? Environmental Conditions vs. Organic Farming Development in the EU with Particular Focus on Poland

Mariusz Malinowski [1,*], Luboš Smutka [2] and Arkadiusz Sadowski [3]

1 Department of Economics, Faculty of Economics, Poznan University of Life Sciences, 60-637 Poznan, Poland
2 Department of Trade and Finance, Czech University of Life Sciences Prague, 165 00 Praha, Czech Republic; smutka@pef.czu.cz
3 Department of Economics and Economic Policy in Agribusiness, Faculty of Economics, Poznan University of Life Sciences, 60-637 Poznan, Poland; arkadiusz.sadowski@up.poznan.pl
* Correspondence: mariusz.malinowski@up.poznan.pl

Abstract: Organic farming takes on particular importance in the context of implementing the sustainable development concept as it combines environmentally safe farming methods with (as a general assumption) producing pollution-free food. Hence, environmental conditions might play a role in determining the development pace of that type of farming. The key objective of this paper is therefore to identify the scope and direction of multidimensional relationships between the development level of organic farming and environmental conditions. This was performed with the canonical analysis. The research process included the structuring of the authors' own synthetic metrics used in assessing the condition of the environment and the development level of organic farming. The study covered European Union countries and all 380 Polish districts (Poland is one of the very few Union members where organic farming development is currently inconsistent with the expected trends adopted under the Common Agricultural Policy). It follows from the analyses that when the variables relating to environmental conditions are known, they can explain only less than 10% of variance in the set of variables used in describing the development level of Polish organic farming. In turn, the analysis at Union level suggests that a positive—but not stronger than moderate—correlation exists between the two phenomena.

Keywords: organic farming; natural environment; pollution; green deal; canonical analysis

1. Introduction

One of the major challenges faced by humanity today is to stop degrading the natural environment [1]. The green transformation of the Union's agricultural policy has been progressing for a few decades now; as a consequence, steps are taken to consistently emphasize the importance of environmental conditions for those measures. This somehow made it easier to come out of the shadow for organic farming which—despite having a century of history—has been marginalized until very recently (for more information, see [2–5]). Even as late as in 1990, former European eastern bloc members (such as Poland) were faced with a chronic lack of goods in the market, and therefore their priority was to maximize production volumes (rather than to improve production quality) while neglecting the environmental aspects.

Despite strong efforts being exercised by Union institutions (including the implementation of Agenda 2000 of 1997 [6], the European Green Deal in 2019 [7] and its component, the "from farm to fork" strategy of May 2020), the development level of organic farming continues to noticeably differ between EU countries. Therefore, it becomes important to investigate and identify the development drivers of organic farming and explore the reasons behind its geographic variation. How much can it be impacted by the quality of the natural environment? The environment and its protection have become a topic of interest to new institutional economics because it may be viewed as a specific public good. The generation of production- and consumption-related externalities by public goods results in delivering benefits to society, which is an aspect of fundamental importance from the perspective of sustainable development economics [8].

The farmers are particularly severely affected by the consequences of climate and environmental change, which has grown in intensity over the last decades. This is what makes "organic farmers the pioneers of future sustainable agriculture" [9]. Therefore, the question arises: can products originating from areas where environmental conditions are generally viewed as poor be referred to as "organic" (despite the certification and control processes)? Do the environmental enhancements go hand in hand with the development of organic farming in the European Union? How does this process look in Poland, one of the very few Union members where organic farming development has, over the last decade, failed to meet the expected trends adopted under the Common Agricultural Policy? Indeed, organic food means, by definition, not only food without harmful residue of pesticides and plant protection products but also food that demonstrates a smaller content of harmful substances—irrespective of whether they result from farming methods or from the condition of the natural environment.

The analyses found in the literature on the subject mostly focus on emphasizing that organic farming has more favorable environmental impacts than conventional agriculture (for instance: [10–12]). It is much less common for the authors to address that problem from the other side of the relationship between these categories, i.e., the impact of the natural environment on the development level of organic farming (which somehow suggests total reliance on the professionalism and trustworthiness of inspection and certification authorities). Therefore, this paper makes an attempt to bridge that research gap, at least partially.

Just like many other Union projects, the implementation of the European Green Deal and a fair green transformation are "knowledge-intensive" domains that require research to be conducted across many fields of science [13]. The purpose of this paper is to identify the multidimensional relationships between environmental conditions and the development of organic farming in the European Union, with a particular focus on Poland. Because of the multifaceted nature of the phenomena considered, the study mostly relied on canonical analysis, a sophisticated method for statistical data exploration. Proposing the use of canonical analysis as a tool for identifying the relationships between multidimensional categories is the applicative purpose of this paper. The analyses also included the structuring of the authors' own synthetic metrics for the phenomena covered (based on the TOPSIS method). The study is of a descriptive and analytical nature and uses data for the years 2013 and 2022 retrieved from Eurostat, the Research Institute of Organic Agriculture, the International Federation of Organic Agriculture Movements, the Polish General Inspectorate of Agri-Food Trade Quality, and the Polish Central Statistical Office. The criterion for the selection of variables (and time range) was the completeness, availability, and timeliness of the data. The article has a theoretical and empirical character. The first part of the article presents, among other things, the importance of environmental conditions in the context of creating the development of organic farming. Next, a linear ordering of EU countries was carried out with regard to the level of development of organic farming and environmental conditions. In addition, a canonical analysis was carried out at the level of 380 districts in Poland.

2. Natural Environment vs. Organic Farming: A Theoretical Introduction

Organic farming was empirically proven to have a more beneficial environmental impact than conventional agriculture (for more information, see [10,14,15]). Its positive outcomes include the content of organic matter, the biological activity of soils, and reduced erosion. Another advantage of organic farming over conventional agriculture is the effect it has on ground and surface waters, which is manifested in smaller leaching of nitrates and pesticides. Also, organic production methods seem to be more climate-friendly, for instance, in terms of carbon and ammonia emissions.

When it comes to organic food, assessments and inspections span all manufacturing processes "from farm to fork". As it is unfeasible to provide a clear final evaluation of an organic food product based on chemical analyses, the consumers are provided with information on natural and controlled processes used in manufacturing it. Organic farming relies on the natural qualities of processes taking place within the farm. It can only be practiced in a previously uncontaminated environment that meets all standards applicable to the presence of substances harmful to health [16]. However, in order to face growing competition, organic farms must be guided by economic criteria that may contradict the farmers' environmental commitment to use natural resources so as not to create an environmental imbalance [17]. Over the last decades, especially in countries with a poorly organized market, such as Poland, an important role has been played by subsidies granted under agri-environmental programs in order to improve the quality of the natural environment in rural areas by supporting environmentally friendly agricultural production methods [18,19]. Farmers who decide to continue or shift to organic production may count on organic payments allocated in the function of the area of organically farmed land. The payment rates for 2023–2027 differ between crop groups and depend on whether the farm is undergoing or has completed the conversion period. Additionally, they are 35% higher than those set in 2021 on average [20].

The literature on the subject presents a number of typologies for drivers of the development of organic farming (in this paper, organic farming development is equated with positive quantitative changes in the number of organic farms, area of organic farmland, production volumes, number of processors, etc.). Although, in addition to financial and institutional aspects, emphasis is often placed on environmental factors (Table 1), the literature generally focuses on the farmers' attitudes towards the natural environment and their willingness to follow environmentally sound farming practices (for more information, see [21,22]) rather than on the quality of the natural environment. However, field plant production is a process that takes part in, and has a direct impact on the environment. Environmental characteristics shape the mix of plant species and varieties, as well as the sensory values and (chemical and biological) pollution levels of agricultural products. Unfortunately, nowadays, land intended for agricultural uses often comes from the conversion of naturally valuable areas such as rainforests [23,24].

This synthetic overview of the sets of variables used in both Polish and foreign literature shows how different (in number and nature) variables are used to assess the level of development of organic farming and how environmental aspects are treated differently from farmers' beliefs about the environment and soil fertility to fertilizer use. The selection of variables is often subordinated to the purpose of the research carried out. The literature review carried out (which is part of the substantive criterion taken into account in the selection of variables) and the availability, completeness, and timeliness of the data, influenced the shape of the sets used by the authors (see Section 3).

Table 1. Development drivers of organic farming.

Author	Development Drivers of Organic Farming in Poland
[25]	Factors that make production economically viable or unviable (e.g., price, margin, subsidy rate); links between farms and the market; distribution forms (related to sales opportunities); institutional solutions for vertical integration of producers; other lines of farm business in addition to food production.
Siedlecka (2015) [26]	Financial (related to the eligibility for financial support); environmental (related to biodiversity, soil fertility, etc.); market-related (resulting from price trends); social (resulting from changing lifestyles); regional (resulting from the agrarian structure, the region being an industrial or agricultural one, labor resources).
[27]	Legal and organizational (e.g., efficiency of certifying authorities); economic (including the average monthly remuneration, share of food spending, farm support); environmental and production-related (including the share of permanent pasture, consumption of mineral fertilizers); social (including agricultural employees per 100 ha).

Table 1. *Cont.*

Author	Development Drivers of Organic Farming in Poland
[28]	Support policy for organic production and food market; seizing export opportunities; combining organic production with the development of agritourism; development of integrators of dispersed production activities and distributors, including producer groups; promoting organic products.
[29]	High support for organic farms; environmental protection; interest in organic farming; family health; low production costs; environmentally friendly production processes; green lifestyle and philosophy; land and labor resources; concern for animal welfare; high price premiums for organic food.
Author	Development Drivers of Organic Farming Around the World
[21]	Farmer's characteristics (education, age, experience, etc.); farm structure (location, size, soil type, etc.); farm management (use of productive inputs, crop diversification, etc.); exogenous factors (production and input prices, market size, subsidies, transformation costs, etc.); attitudes and opinions (farmers' views on the environment, acceptance from the rural community, lifestyle, health and environmental protection concerns, etc.).
[30]	Availability of information and knowledge; economic and financial motives; technical and managerial skills; social aspects; environmental concerns; institutional environment; demographic factors.
[22]	Natural capital factors (including the farm's geographic location, amount of spraying, motivation to maintain the environment in a good condition); human capital factors (including a personal approach to organic products, education level); social capital factors (including supportive social networks, certification); financial capital factors (including economic benefits, additional margin); physical capital factors (including farmland area, certification legislation).
[31]	Environmental protection factors; innovativeness factors; economic factors; social factors; health factors.
[32]	Exogenous factors (demand for organic products, price, market access, available technologies, education, knowledge transfer, partner networks; social attitudes and subsidies); endogenous factors (location, farm size, expected costs, benefits, knowledge, use of information and telecommunication technologies, farmers' age, education, gender, off-farm activity, attitudes and beliefs related to organic farming, and willingness to protect the environment).
[33]	Organic heritage; farm size; primary agricultural sector; market for organic products; subsidies; and other economic factors affecting the conversion into organic farming.
[34]	Effective institutional leadership; affordable third-party certification; phased reduction of agrochemicals through clear political pathways; development of extension networks; market access with price premiums; the provision of organic inputs.

3. Materials and Methods

Works by R.M. Solow and J.E. Meade began the development of the neoclassical growth theory, which included analyzing the relationships between environmental pollution and economic growth. While this approach pays particular attention to microeconomic processes, the predominant idea of Keynesian economics is that intergenerational environmental fairness is of fundamental importance to the management of natural resources. Also, Keynesian environmental economics emphasizes that neoclassical economics fails to fully address the ecological problems it takes into consideration. Therefore, it has become meaningful to seek new methods for analyzing ecological problems in order to better understand the essence of the interrelations between nature and economics [14]. The canonical analysis, an approach quite rarely used in social sciences, could be one of them. Considering the multifaceted nature of the phenomena covered by this study, it seems reasonable to use that very multidimensional exploration technique in assessing these interactions. In this case, using tools such as multiple regression models and analyzing each dependent variable separately could contribute to the possibility of narrowing the results of analyses because there would be a risk of losing relevant information on interactions in the set of explained variables.

The canonical analysis of data for Poland was preceded by a classical correlation analysis, and by structuring a synthetic metric used in assessing the condition of the EU's natural environment. It was quantified with a metric developed by the authors based on the TOPSIS method (Technique for Order Preference by Similarity to an Ideal Solution) proposed by C.L. Hwang and K. Yoon (for more information, see [35–37]). The correlation analysis performed is probably one of the most popular research tools regardless of scientific discipline. It is important to be aware that it is used to detect relationships between two characteristics, but it should be remembered that it does not take into account which variable is the cause and which is the effect.

Sub-variables (both for the European Union as a whole and for Poland) were selected in two steps. Initially selected were those which, according to the authors' know-how, are of great importance in the context of quantifying the phenomena under consideration. At this stage, the selection criterion was the availability of up-to-date and complete data for all objects. It was assumed that the variables covered by the study would be expressed as indexes (for instance, per population or per unit of physical area) rather than as absolute values.

Initially, a set of 16 variables was proposed to be used in structuring the synthetic environmental metric (national environmental index) for EU countries: NEI1: percentage of forested areas; NEI2: percentage of Natura 2000 protected areas; NEI3: carbon dioxide emissions per capita; NEI4: methane emissions per capita; NEI5: net greenhouse gas emissions per capita; NEI6: generation of hazardous and non-hazardous waste per capita; NEI7: renewable freshwater resources per capita; NEI8: Water Exploitation Index plus (percentage use of total renewable freshwater resources available in the river basin in the reference period); NEI9: carbon footprint of "crop and animal production, hunting and related service activities" per capita; NEI10: material footprint per capita (the metric for the amount of raw materials required across the supply chain to meet the final demand for goods); NEI11: share of renewable energies; NEI12: household energy consumption per capita; NEI13: municipal waste recycling ratio [%]; NEI14: packaging waste recycling ratio [%]; NEI15: nitrogenous fertilizer consumption in tons per hectare; NEI16: phosphorus fertilizer consumption in tons per hectare.

In the second stage of the selection of variables, statistical criteria (degree of variation and mutual correlation) were used in order to narrow the original group of variables in both sets (describing the environmental conditions and the level of development of organic farming). An assumption was adopted that the characteristics with a coefficient of variation below a critical threshold value (set arbitrarily at a level of 10% for 2013 and 2022) would be eliminated. Based on that criterion, all variables were subject to further analysis. Another important criterion for the selection of variables is their mutual correlation (the capacity criterion). As two highly correlated variables deliver similar information, it is recommended to eliminate one of them. A method referred to as the inverse correlation matrix was used to verify the information value (for more information, see [38]). Where necessary, the variable with the highest diagonal entry (above the threshold set arbitrarily at 10) was eliminated as the next step. This was the basis for removing the variable relating to greenhouse gas emissions (as the only ratio that exceeded the threshold value in both periods).

A set of 31 variables was preliminarily proposed to describe the development level of organic farming in Poland: A1: number of organic farms per capita; A2: number of processing plants per capita; A3: organic farm area per capita; A4: area of land under cereals per capita; A5: cereal production volume per capita; A6: area of land under legumes grown for dry seed per capita; A7: production volume of legumes grown for dry seed per capita; A8: area of land under potatoes per capita; A9: potato production volume per capita; A10: area of land under beet and root crops per capita; A11: beet and root crops production volume per capita; A12: area of land under industrial crops per capita; A13: industrial crops production volume per capita; A14: area of land under fiber plants per capita; A15: fiber plants production volume per capita; A16: area of land under vegetables per capita; A17: vegetable production volume per capita; A18: area of land under fruit plants and berries per capita; A19: fruit plant and berry production volume per capita; A20: area of land under fodder plants; A21: fodder plant production volume per capita; A22: area of meadows and pastures per capita; A23: cattle population per capita; A24: pig population per capita; A25: ovine population per capita; A26: caprine population per capita; A27: poultry population per capita; A28: equine population per capita; A29: rabbit population per capita; A30: milk production volume per capita; A31: egg production volume per capita.

In turn, the following 24 variables were proposed to be used in determining the environmental conditions: S1: water consumption per capita; S2: share of the industrial sector in water consumption; S3: population served by wastewater treatment plants as a percentage of total population; S4: treated municipal and industrial wastewater as a percentage of wastewater which requires treatment; S5: total capacity of wastewater treatment plants per 1000 population; S6: total capacity of wastewater treatment plants with enhanced biological nutrient removal per 1000 population; S7: sediments from industrial wastewater treatment plants generated within a year per 1000 population; S8: biochemical oxygen demand of treated wastewater per 1000 population (an index describing the demand for oxygen necessary to oxidize organic compounds present in wastewater in aerobic conditions); S9: chemical oxygen demand of treated wastewater per 1000 population (amount of oxygen necessary to oxidize organic and inorganic compounds present in wastewater); S10: total suspended sediments in treated wastewater per 1000 population; S11: total chloride and sulfate ions discharged into water or soil per 1000 population; S12: total nitrogen in wastewater discharged into water or soil per 1000 population; S13: total phosphorus in wastewater discharged into water or soil per 1000 population; S14: total gaseous pollutant emissions from particularly noxious plants per sq. km; S15: sulfur dioxide emissions from particularly noxious plants per sq. km; S16: carbon oxide emissions from particularly noxious plants per sq. km; S17: nitrogen oxide emissions from particularly noxious plants per sq. km; S18: total particulate matter emissions from particularly noxious plants per sq. km; S19: particulate matter emissions retained in or neutralized by pollutant reduction systems as a percentage of generated pollutant emissions; S20: difference between the number of planted and felled trees per 100 sq. km; S21: parks, greenways and housing estate greenery per 100 sq. km; S22: legally protected areas per 100 sq. km; S23: number of active landfill sites used in waste neutralization per 1000 population; S24: area of active landfill sites used in waste neutralization per 1000 population.

The analysis of variation of variables in both sets did not result in eliminating any of them (the classic coefficient of variation was above 10% of each variable). Conversely, due to a high correlation level, variable A20 was eliminated from the set relating to the development level of organic farming following the assessment of the information potential (the diagonal entry of the inverse correlation matrix was above 10). In the second set covered by the study, that criterion served as a basis for eliminating variables S6, S11, and S20.

A canonical analysis was performed to present the multidimensional dependencies between the sets of variables proxying for the development level of organic farming and environmental conditions in Poland. The primary datasets were statistically validated in a way similar to that employed when structuring the synthetic environmental metric for the European Union as a whole (by taking the variation and correlation levels into account).

The canonical analysis used in this study means multiple linear regression generalized for two sets of variables. In this case, the exploration of relationships between two datasets boils down to investigating the links between two new types of variables. Referred to as canonical variables, they are calculated as weighted sums of the first and second set of variables (explained and explanatory variables). The weights are selected so that the two weighted sums are maximally correlated with each other [39–43]. When considering two linear combinations, $x = x^T\hat{w}_x$ and $y = y^T\hat{w}_y$, the objective is to maximize the following expression:

$$r_l = \frac{\left(w_x^T R_{xy} w_y\right)}{\sqrt{\left(w_x^T R_{xx} w_x w_y^T R_{yy} w_y\right)}}, \tag{1}$$

where: R_{xx}: correlation matrix for explained variables (relating to environmental conditions); R_{yy}: correlation matrix for explanatory variables (relating to the development level of organic farming); R_{xy}: correlation matrix for both types of variables; w_x, w_y: weights for first-type and second-type canonical variates.

It should be mentioned at this point that canonical analysis is sensitive to outliers. When testing the relationship between canonical variables, conditions as to the normality of variable distributions should be met (which is difficult in economic sciences). Obtaining reliable results from the analysis requires a sufficiently large sample size (at least 50 observations).

The statistical significance of canonical variates generated in this study was verified with the Wilks' lambda test (cf. [44,45]). The test statistic for a set of *s-k* variables was used to verify the significance of pairs of canonical variates:

$$\Lambda_k = \prod_{l=k}^{s}\left(1 - r_l^2\right), \tag{2}$$

where: *s*: number of canonical variates.

The results of a canonical analysis are sensitive to outliers. They were identified with the modified three-sigma rule (cf. [46]) which consists of removing the observations that fail to meet the following condition:

$$\frac{x_{i-M}}{MAD} > |\pm 3|, \tag{3}$$

where: *MAD*: mean absolute deviation, *M*: median, x_i: value of the characteristic.

If identified, outliers were replaced with the median calculated for regions (NUTS 2), which are home to districts with sub-variables outside the defined thresholds. In the set of variables relating to the development level of organic farming, it was needed 30 times (each time because it exceeded the upper boundary of the acceptable interval). In turn, as regards environmental conditions, it was the case 24 times (22 and 2 times due to exceeding the upper and lower boundary, respectively, of the acceptable interval).

All variables subject to the canonical analysis should follow a normal distribution. The results of the Shapiro–Wilk test were used to assess whether this was the case. The variables which did not follow a normal distribution were subject to the Box–Cox transformation [47] so as to make their distribution as close as possible to normal (the transformation parameter was selected based on the maximum likelihood estimation from a customarily defined interval of $[-5, 5]$).

The canonical analysis was performed at the level of all 380 Polish districts (in accordance with the Nomenclature of Territorial Units for Statistics, Polish districts are LAU 1 local units). The analysis was carried out at the district level (rather than NUTS 2 regions), which allowed an increase in the number of objects covered by the study and to extend the number of potential diagnostic variables (compared to a LAU 2-level analysis). However, due to an insufficient number of objects (countries), the decision was made not to perform a similar analysis for the whole EU.

The methodology used in the study is presented by means of a flow chart (Figure 1). The arrows mark the transition to the next step in the methodological procedure.

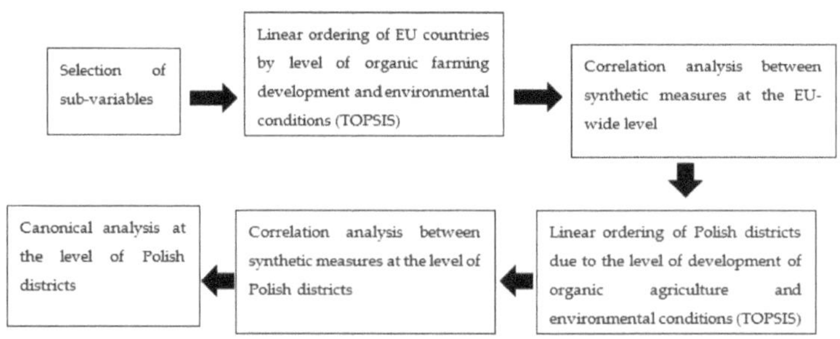

Figure 1. Flowchart of the methodology.

4. Results

4.1. Synthetic Assessment of the Level of Development of Organic Farming and the State of the Environment in the European Union

The European Union is among the key players involved in the development of organic farming. In accordance with Eurostat data, there were over 14.7 million hectares (ca. 9.1%

of the Union's total agricultural land) of organic farmland in the EU in 2020, compared to slightly above 9.5 million hectares (5.9% of agricultural land in the EU-27) in 2013. Over the study period (2013–2022), most Union countries (except for Poland) witnessed a considerable increase in the area of land under organic crops (cf. Table 2). Also, in most countries (except for Poland, Romania, and Sweden), that process was accompanied by an increase in the number of organic producers. The highest number of organic farmers per 1000 population was recorded in Greece (3.45), Austria (2.88) and Latvia (2.22). Conversely, the smallest ratios were reported in Malta (0.05), the Netherlands (0.11) and Belgium (0.22).

Table 2. Organic farming in European Union countries in the context of the synthetic metric of the condition of the natural environment.

	Number of Organic Producers Per 1000 Population			Share of Organic Farmland in Total Area of Agricultural Land [%]			Share of Permanent Pasture in Organic Farmland in 2020 [%]	Ratio Between the Number of Producers and the Number of Processors in 2021	SEM		SMOFD	
	2013	2022 *	Percent Growth Between 2013 and 2022 [%]	2013	2021 **	Absolute Growth Between 2013 and 2021			2013	2022	2013	2021/2022
BE	0.15	0.22	51.53	4.67	7.48	2.81	62.22	1.63	0.53	0.53	0.27	0.19
BG	0.53	0.66	24.21	1.13	1.71	0.58	25.94	23.86	0.62	0.58	0.30	0.35
CZ	0.37	0.47	25.47	13.47	15.55	2.08	81.87	5.16	0.58	0.57	0.31	0.27
DK	0.46	0.69	51.55	6.44	11.58	5.14	15.98	3.60	0.54	0.53	0.40	0.39
DE	0.29	0.43	50.94	6.04	9.65	3.61	52.43	1.86	0.59	0.62	0.36	0.25
EE	1.18	1.50	26.92	15.65	22.97	7.32	42.53	10.48	0.57	0.63	0.43	0.49
IE	0.29	0.42	44.93	1.20	2.00	0.80	89.05	8.90	0.5	0.47	0.10	0.09
GR	2.01	3.45	71.24	7.36	10.15	2.79	54.99	18.07	0.62	0.62	0.42	0.51
ES	0.66	1.16	77.21	6.85	10.79	3.94	52.23	8.93	0.65	0.67	0.31	0.32
FR	0.38	0.54	40.04	3.66	9.67	6.01	34.93	3.02	0.59	0.59	0.34	0.31
HR	0.38	1.59	320.55	3.13	8.26	5.13	38.98	15.94	0.67	0.65	0.30	0.39
IT	0.76	1.29	70.49	10.6	16.83	6.23	27.87	3.19	0.64	0.64	0.43	0.43
CY	0.87	1.41	62.02	4.03	6.43	2.40	3.14	18.46	0.53	0.52	0.40	0.46
LV	1.74	2.22	27.03	9.89	15.34	5.45	45.56	64.17	0.67	0.68	0.43	0.66
LT	0.87	0.93	6.02	5.74	8.91	3.17	34.88	20.90	0.61	0.63	0.35	0.38
LU	0.15	0.23	49.33	3.39	5.19	1.80	51.16	1.15	0.55	0.59	0.34	0.20
HU	0.17	0.64	278.70	2.45	5.81	3.36	60.03	10.49	0.59	0.57	0.23	0.22
MT	0.02	0.05	120.06	0.06	0.61	0.55	0.00	1.56	0.56	0.55	0.36	0.35
NL	0.10	0.11	13.67	2.65	4.22	1.57	58.01	1.99	0.54	0.56	0.46	0.17
AT	2.57	2.88	12.20	18.4	25.69	7.29	57.71	14.17	0.67	0.66	0.52	0.57
PL	0.70	0.54	−22.27	4.65	3.78	−0.87	16.84	27.84	0.51	0.53	0.32	0.40
PT	0.29	1.30	346.39	5.31	19.31	14.00	61.50	10.23	0.67	0.66	0.18	0.40
RO	0.73	0.61	−16.82	2.06	4.42	2.36	33.07	55.32	0.60	0.54	0.29	0.47
SI	1.48	1.76	18.89	8.07	10.81	2.74	80.10	26.51	0.72	0.71	0.32	0.38
SK	0.06	0.30	375.55	8.18	13.45	5.27	65.15	6.02	0.65	0.66	0.25	0.26
FI	0.79	0.89	13.09	9.07	14.45	5.38	0.60	12.09	0.60	0.61	0.45	0.47
SE	0.58	0.48	−16.62	16.5	20.20	3.70	22.67	4.75	0.71	0.68	0.47	0.43

Symbols: * if no data were available for 2022, data for 2021 are presented (for France and Latvia, the most recent data come from 2017 and 2019, respectively); ** if no data were available for 2021, data for 2020 are presented. SEM: synthetic environmental metric; SMOFD: synthetic metric of organic farming development (value aggregated using the TOPSIS method for the four variables [1, 2, 3, 4] presented in the table).

The development level of organic farming in the European Union is largely related to environmental conditions specific to each country. Countries affected by less favorable natural conditions for agricultural production (Austria, Sweden, Italy) demonstrate a greater share of land under organic crops in their total area of agricultural land (as the production capacity is limited, the financial support system for organic farming becomes a way to improve the farmers' financial situation). Conversely, in countries with more favorable natural conditions (including due to soil types and water availability), such as France and the Netherlands, the share of organic farming is relatively small (9.7% and 4.2%, respectively). The share of organic farmland in total agricultural land exceeds 20% in only three countries: Austria (which also has the second highest value of the synthetic metric of organic farming development), Estonia (ranked 4th), and Sweden (ranked 9th). The above means that despite the Union's numerous transition incentives, organic farming did not become a widespread production system. In 13 countries, permanent pasture (rather than land intended for sowing or permanent crops) accounts for most of the area of organic farmland. The largest share of permanent pasture in organic farmland was recorded in Ireland and the Czech Republic (89% and 82%, respectively). At the EU level, permanent pasture represents more than 42% of organic farmland.

Over the study period, the greatest percent increase in the share of organic farmland in the total area of agricultural land was recorded in Portugal (by 14 percentage points) and in Estonia and Austria (ca. 7 percentage points). Poland was the only Union country to witness a decline in that ratio (by 0.9 percentage points). Over the study period, in the group of countries that demonstrated a small share (below 4%) of organic farmland in 2013, the greatest growth was recorded in France (6.01 percentage points) and Croatia (5.13 percentage points). The increase in the share of organic farmland is quite commonly equated with being provided with public support which is a strong incentive to switch from conventional to organic farming.

In Poland, this is all the more surprising since it offers great development opportunities for organic farming. Therefore, the share of organic farmland was forecasted to sharply grow to as much as 10–15% of the area of agricultural land in the first years following the accession to the EU. The aspects that were supposed to make it happen included [18,20] traditional farming being dominated by small and medium-sized family farms and a high share of agricultural employment; relatively low levels of environmental pollution; and price competitiveness of organic products. Moreover, the prevalence of lowland areas, a moderate climate, and the availability of low- and medium-quality soils that respond well to organic fertilizers were also viewed as an advantage. In turn, the main limitations faced by organic production in Poland include a poorly organized market and an inefficient distribution network, as well as low-income levels of a large part of consumers, which reduces their purchasing power with respect to more expensive products. In Poland, the number of operators who accessed financial support for organic farming has reduced over the recent years, mostly because of their failure to meet the applicable legal regulations and due to a heavy bureaucratic burden. This is especially true for farmers who do not combine plant and animal production and thus fail to comply with the requirement for minimum livestock numbers. Another important problem is the instability of legal regulations that govern organic farming aspects, which makes the decision to convert a riskier process.

Polish organic farming is strictly related to public support; without subsidies, it would either struggle to survive or be only a niche activity for a small group of farmers, as was the case before joining the EU. As demonstrated by A. Sadowski et al. [48], without subsidies, Polish organic farms could only generate a fraction of the surplus earned by their conventional peers. The share of subsidies in net value added is much greater in organic farms; between 2016 and 2018, it accounted for 76%, which makes Polish organic farms virtually totally dependent on public aid.

Based on Eurostat data [49], it can be assumed that the Polish organic farming sector includes a relatively small number of farms covered by support and by the certification system per capita (ranked 10th according to 2022 data) while also having an extremely small number of processing plants (which is a disadvantageous situation). In Poland, the ratio between the number of organic producers and that of processors was 28; only Latvia (64) and Romania (55) reported an even worse proportion (the former having, however, the highest aggregated metric of the organic farming development level). For comparison, Poland's southern neighbors, the Czech Republic and Slovakia, recorded a ratio of 5 and 6, respectively.

These values are somehow reflected in the value of organic products sold. In 2021, European Union residents spent an average of EUR 104.3 (with the highest amounts being recorded in Switzerland, Denmark, and Luxembourg: EUR 424, EUR 384, and EUR 313, respectively) on organic goods [50]. For comparison, as regards Visegrad Group countries, per-capita spending on organic products was EUR 21.9 in the Czech Republic, EUR 8.3 in Poland, and EUR 3.0 in Hungary. In Slovakia, it was EUR 1 [51] according to data published in 2021 (although it is consistent with the amount recorded in 2010).

S. Heinze and A. Vogel [52] believe that if the objective of a national agricultural policy is to increase the share of organic farmland, it is important not only to encourage conventional farmers to engage in organic farming but also to prevent organic farmers from quitting the organic sector.

The location of an organic agricultural producer should take into account the possible threats, if any, caused by different kinds of environmental pollution. Therefore, that type of farming should reach the highest development levels in regions where industrialization is (relatively) low, whereas natural values (in both quantitative and qualitative terms) are abundant. It follows from the analysis carried out at the European Union level that the relationship between the condition of the natural environment (quantified using the authors' own synthetic metric) and the number of organic producers, the share of organic farmland in total area of agricultural land, or the aggregated metric of the development level of organic farming is moderate, if not weaker (is statistically significant at $p < 0.05$) (Table 3).

Table 3. Correlation (Pearson's coefficient).

Specification	2013	2022
SEM vs. the number of organic producers	0.4160	0.4818
SEM vs. the share of organic farmland in total area of agricultural land	0.4604	0.6444
SEM vs. SMOFD	0.1937	0.4552

Whether for the aggregated metric of agricultural development level, the number of organic producers, or the area of land under organic crops, the correlation was stronger in the last year of the study period. However, it would be difficult to assert that environmental enhancements go everywhere hand in hand with the development of organic farming in each European Union country.

4.2. Relationships Between the Development Level of Organic Farming and the Condition of the Natural Environment in Poland

The results of the linear ordering of Polish districts by development level of organic farming and by the environmental conditions score (quantified using the TOPSIS method) are presented on a percentile map (Figure 2). The figure was supplemented with the list of top 10 and bottom 10 synthetic metric values and with selected dispersion measures.

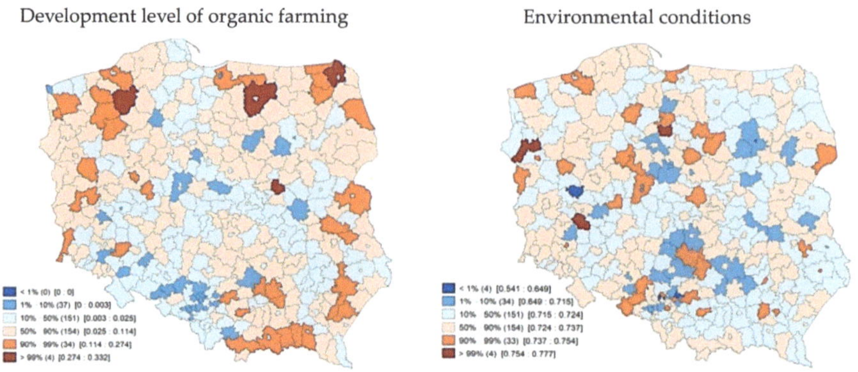

District	Synthetic metric	District	Synthetic metric
Szczecinek	0.3315	Myślibórz	0.7774
Suwałki	0.3066	Świętochłowice	0.7574
Olsztyn	0.2954	Wąbrzeźno	0.7566
Warsaw	0.2772	Góra	0.7541
Olecko	0.2649	Nowy Dwór Mazowiecki	0.7528
Przeworsk	0.2533	Grudziądz	0.7502
Gołdap	0.2418	Radomsko	0.7499
Łęczna	0.2350	Olkusz	0.7495
Krosno	0.2216	Międzychód	0.7492
Środa Śląska	0.2109	Rzeszów	0.7492
	...		
Dąbrowa Górnicza	0.0004	Chełm	0.6985
Bytom	0.0004	Konin	0.6978
Siemianowice Śląskie	0.0003	Jaworzno	0.6956
Zabrze	0.0003	Toruń	0.6937
Chorzów	0.0002	Krapkowice	0.6920
Głogów	0.0001	Puławy	0.6875
Kutno	0.0001	Bełchatów	0.6811
Garwolin	0.0001	Płock	0.6718
Rybnik	0.0001	Lubliniec	0.6625
Jawor	0.0000	Kwidzyn	0.6565
	Variation		
AM	0.0468	AM	0.7155
Vs [%]	123.3117	Vs [%]	2.3935
SD	0.0577	SD	0.0171
MED	0.0251	MED	0.7160
Q1	0.0082	Q1	0.7113
Q3	0.0621	Q3	0.7226

Figure 2. Percentile map for the synthetic metric of development level of organic farming and environmental conditions, together with the list of top 10 and bottom 10 values. Symbols: AM: arithmetic mean, Vs: classic coefficient of variation, MED: median, Q1: first quartile, Q3: third quartile. Source: own study based on the Local Data Bank of the Central Statistical Office [53] and on the database of the Inspectorate of Marketable Quality of Agri-food Products [54].

It is quite rare for the districts to demonstrate high levels (above the 90th percentile) of both the development of organic farming and the environmental conditions score. In the set of variables used in this study, the highest values of the synthetic metric of organic farming development were identified in the following districts: Szczecinek (Zachodniopomorskie voivodeship), Suwałki (Podlaskie voivodeship) and Olsztyn (Warmińsko–Mazurskie voivodeship). These regions were found to have relatively high levels of variables relating to aspects such as the number of producers, farm area, area of land under vegetables and cereals, and bovine numbers. Also, it is interesting to see that the capital district of Warsaw scores highly in the group of districts with the highest development level of organic farming (usually, organic farming is a rural activity). The Mazowieckie voivodeship (where Warsaw is located) has the largest area of agricultural land in the country (nearly 2.5 million ha, i.e., almost 13% of total agricultural land in Poland); in it, organic farmland accounts for over 41,000 hectares (7.4% of total organic farmland in Poland). Also, according to 2022 data from the Inspectorate of Marketable Quality of Agri-food Products, the voivodeship is home to 2873 organic producers (12.56% on a countrywide basis) who have their registered offices in the capital city of Warsaw (and thus overstate the statistics of this large urban agglomeration). The increase in the number of organic farmers in the capital city is driven by the presence of an absorptive market (the Warsaw agglomeration, i.e., the capital and its suburbs, has a population of ca. 3 million) and of wealthy consumers. Note also that 9 out of the 13 Polish certification authorities are based in Warsaw. Most (8) of the 15 districts with the lowest values of the synthetic metric are cities. Also, 8 out of the bottom 15 districts (in terms of organic farming development) are located in the Śląskie voivodeship. It is the most urbanized and most densely populated Polish region whose central part is home to the Upper Silesian Industrial Region (the most industrialized area in Poland). These are the areas with extremely low values of particular sub-variables (usually the lowest ones across the country). The variation in development levels of organic farming across the Polish territory can be viewed as extremely high. The coefficient of variation was more than 123%, with the mean value of the synthetic metric at 0.046, a minimum of 0.00, and a maximum of 0.33.

Conversely, the variation in the synthetic metric of environmental conditions was identified to be considerably smaller. The coefficient of variation was 2.4%, and the values of the metric oscillated between 0.55 and 0.76. The units with the highest values of the metric structured in this study were found to have small or extremely small values of sub-variables related to aspects such as the level of biochemical and chemical oxygen demand or nitrogen and phosphorus content in wastewater (viewed as having an inhibiting effect) while reporting relatively high levels of variables related to the share of retained or neutralized particulate matter emissions or the area of landfill sites used in waste neutralization. As regards the bottom-ranked districts in terms of environmental conditions, variables with low values were mostly those relating to the surface of legally protected areas, and the small number and area of landfill sites. In turn, those having an inhibiting effect (sulfur, nitrogen, and carbon oxide emissions) reached high levels.

The correlation analysis carried out exclusively with data for 380 Polish districts discovered a positive yet statistically insignificant correlation (at the significance level of 0.05) between the synthetic metric of the development level of organic farming and the synthetic metric of environmental conditions (measured with TOPIS metrics). The Pearson correlation coefficient was 0.0583.

4.3. Canonical Analysis

The input dataset served as the basis for generating 21 canonical roots, which correspond to the number of variables covered by a reduced set relating to environmental conditions (Table 4). The first pair of canonical roots explains most relationships between the sets under consideration, and therefore research practice places the greatest emphasis on the correlation between them; of all estimated correlation coefficients, the first one corresponds to the maximum correlation between the combinations of dependent variables (representing the development level of organic farming in Poland) and independent variables (used in describing the environmental conditions). However, the first pair of the whole set of canonical variates fails to fully explain the relationships, and therefore, it becomes meaningful to determine successive pairs of canonical roots that explain the relationships in less significant dimensions. The canonical variates generated this way are non-correlated and explain increasingly smaller amounts of variation. Hence, the analysis was carried out for all statistically significant canonical roots.

Table 4. Removing successive roots based on the results of the Wilks' lambda test.

Removed Root	Canonical Correlation (R)	χ^2 Test Value	Number of Degrees of Freedom for the χ^2 Test	p Likelihood for the χ^2 Test	Wilks' Lambda Statistic
0	0.5750	792.6901	630	0.0000	0.1059
1	0.5138	651.0057	580	0.0225	0.1582
2	0.4771	542.8281	532	0.3643	0.2149
3	0.4515	451.6721	486	0.8641	0.2782
4	0.4197	371.2011	442	0.9935	0.3494
5	0.3622	302.8202	400	0.9999	0.4241
6	0.3380	253.1669	360	1.0000	0.4881
7	0.3282	210.3386	322	1.0000	0.5511
8	0.2795	170.1071	286	1.0000	0.6176
9	0.2668	141.3870	252	1.0000	0.6700
10	0.2575	115.3151	220	1.0000	0.7213
11	0.2259	91.1054	190	1.0000	0.7725
12	0.2131	72.6204	162	1.0000	0.8141
13	0.1918	56.2217	136	1.0000	0.8528
14	0.1793	42.9858	112	1.0000	0.8854
15	0.1733	31.4469	90	1.0000	0.9148
16	0.1481	20.6777	70	1.0000	0.9431
17	0.1226	12.8499	52	1.0000	0.9643
18	0.1178	7.5034	36	1.0000	0.9790
19	0.0767	2.5682	22	1.0000	0.9928
20	0.0372	0.4875	10	1.0000	0.9986

The highest canonical correlation was nearly R = 0.58, and the value of the Wilks' lambda test used to verify the significance of the highest canonical correlation was 0.1059. This is the correlation between weighted sums in each set, with the weights being calculated for successive canonical variates. Save for the two first canonical variables, other pairs identified are not correlated with each other in a statistically significant way (at $p > 0.05$) and, therefore, (as mentioned earlier) are not covered by the description below.

When analyzing the relationships between multifaceted categories, such as the development level of organic farming and the environmental condition, it is important to explore the structure of dependencies between the defined datasets used in the process. The canonical weights calculated for both sets of variables make it easier to explore the structure of canonical variates by showing the share of each variate in the weighted sum. The weights for standardized datasets are interpreted in a way similar to beta coefficients in the quite popular multiple regression procedure.

As regards the first canonical variate, the highest (absolute) weight values are associated with variables A2 (0.54) and S18 (−0.90), which means that the number of processing plants (per capita) and the amount of particulate matter emissions contributed the most to the first canonical variate. In turn, when it comes to determining the second statistically significant canonical variate using the sub-variables covered by the study, the greatest contribution was recorded for A1 (−1.22) related to the number of organic farms per capita and for S3 (0.89) related to the percentage of the population served by wastewater treatment plants (Table 5).

To dive deeper into analyzing the structure of statistically significant canonical roots, this study also calculated the values of canonical factor loadings, which are equated with coefficients of correlation between a canonical variate and input variables. The higher their absolute value, the greater the importance that should be associated with the variate when interpreting the relationship. In further analyses, the critical value of that correlation coefficient was set at 0.40.

Table 5. Canonical weights and factor loadings.

Variables	Variables Relating to the Development Level of Organic Farming				Variables Relating to Environmental Conditions				
	Canonical Weights		Factor Loadings		Variables	Canonical Weights		Factor Loadings	
	1	2	1	2		1	2	1	2
A1	0.1764 *	−1.2216	0.4037	−0.1005	S1	0.3294	0.3031	0.3706	0.2291
A2	0.5445	−0.2325	0.2646	0.1185	S2	−0.2193	0.2657	−0.0150	0.2746
A3	0.4207 *	1.1234	0.4724	0.2644	S3	−0.4976	0.8858 *	0.0879	0.7170
A4	0.2926 *	0.1013	0.4093	0.2422	S4	−0.0927	0.0747	−0.1852	0.1851
A5	−0.1759	−0.0908	−0.3717	0.1600	S5	−0.0141	0.0489	0.0453	0.2770
A6	0.0856	0.1613	−0.2270	0.2408	S7	0.0178	−0.1041	0.0234	−0.0667
A7	0.1670	−0.2651	−0.1577	0.1119	S8	−0.0698	0.1897	0.2232	−0.0275
A8	−0.1698	−0.0783	−0.2730	−0.2624	S9	0.0789	−0.0246	0.2204	−0.0128
A9	0.3384	0.0420 *	−0.2481	0.4072	S10	0.0199	−0.2183	0.2621	−0.0348
A10	0.0136	−0.0436	0.0590	0.0597	S12	−0.0502	−0.1088	0.2561	−0.0216
A11	0.0461	0.0551	0.1143	0.0659	S13	0.2571	0.0753	0.2581	0.0364
A12	−0.2794	0.0128 *	−0.3462	0.4861	S14	−0.3427 *	−0.6172	−0.5057	0.2219
A13	0.0292	0.2498 *	−0.2334	0.4401	S15	−0.3171 *	−0.4637	−0.4628	0.1874
A14	0.0444	−0.0857	−0.0949	0.0736	S16	−0.2587 *	−0.3553	−0.4158	0.1118
A15	0.0482	−0.0606	−0.0671	−0.0126	S17	−0.1096 *	0.5622	−0.4268	0.2246
A16	−0.2289	0.0517	−0.3019	0.3009	S18	−0.9032	−0.2245	0.0979	0.2007
A17	−0.2443	−0.2118	−0.2265	−0.1470	S19	−0.0468	0.2775	−0.1574	0.3171
A18	0.4623	0.0266	−0.2395	0.0851	S21	0.8614 *	0.3887	0.5447	0.3991
A19	−0.2094	0.1788	−0.1040	−0.0181	S22	0.1626 *	−0.2917	−0.4836	−0.3436
A21	−0.1189	0.3355	−0.3875	0.0603	S23	0.1585	0.2129	−0.0674	0.0500
A22	0.0574 *	0.2136	0.4263	0.2320	S24	−0.0447	−0.2127	−0.0860	−0.0405
A23	0.2067 *	−0.3265	0.6612	−0.1378					
A24	−0.1337	−0.0086	−0.1913	−0.1980					
A25	0.3607 *	0.0825	0.6153	0.0168					
A26	0.0451	−0.0071	−0.3353	0.1116					
A27	0.0000	−0.0294	−0.2571	−0.1628					
A28	0.0267	−0.1661	−0.2099	−0.2000					
A29	0.0010	0.1045	−0.0512	0.0085					
A30	−0.4569 *	−0.1522	−0.6764	−0.3648					
A31	0.0247	−0.3740 *	−0.0328	−0.5200					

Symbols: 1, 2: numbers of statistically significant canonical roots. * values of weights for which the factor loadings are in excess of 0.40.

When it comes to the set of variables relating to the development level of organic farming, the first and the second canonical roots have the greatest factor loadings for the variables A30 (−0.68) and A31 (−0.52), respectively, which relate to the production volume of milk products and eggs. Conversely, in the set of variables relating to environmental conditions, the first canonical root has the greatest factor loading for the variable S21 (0.54); and the second canonical variate has the greatest factor loading for the variable S3 (0.72). They relate, respectively, to the saturation with greenery areas (e.g., parks) and to the percentage of the population served by wastewater treatment plants.

The literature on the subject is not fully consistent on whether the interpretation of existing relationships should be based on factor loadings or canonical weights [cf. [44]]. The use of canonical factor loadings is justified by them being quite intuitive to grasp. However, when adopting this approach, note that they indicate how much correlation there is between single input variables and canonical variates; unlike canonical weights, they do not take account of co-variability effects inside the set of input variables under consideration. Hence, the interpretation of canonical roots based on correlation coefficients can lead to other findings than a more complete "multidimensional" interpretation underpinned by an analysis of canonical weights. This study relied on the second approach.

Based on values of canonical weights and factor loadings, it can be concluded that the first statistically significant canonical root explained the following relationships:

- the greater the saturation with greenery areas (S21) and legally protected areas (S22), the greater the per-capita area of: organic farms (A3); land under cereals (A4); and meadows and pastures (A22);
- the share of greenery areas (S21) and legally protected areas (S22) in the total land area has a positive impact on bovine numbers (A23) and ovine numbers (A25);
- the increase in emissions of: total gaseous pollutants from particularly noxious plants (S14), sulfur dioxide (S15), carbon oxide (S16), and nitrogen oxides (S17) entails a decline in: the number of organic farms (A1), farm area (A3) and area of land under cereals (A4).

In analyzing the factor loadings and canonical weights for the second statistically significant canonical root, it can be noticed that:

- a positive relationship exists between the percentage of the population served by wastewater treatment plants (S3) and the area of land under industrial crops (A12),
- an increase in the percentage of the population served by wastewater treatment plants (S3) entails growth in the production volume of potatoes (A9), industrial plants (A13), and eggs (A31).

By squaring the factor loading values (representing the correlation), the authors determined how much variance of a variable is explained by the canonical variate. If the mean value of that proportion is calculated for all variables, it tells the average percentage of variance explained by the given canonical variate in that dataset. This is referred to as variance extracted (Table 6). In turn, a new "synthetic indicator" referred to as redundancy of a set of variables with respect to another set is obtained by multiplying the eigenvalues (squared values of successive canonical correlations) of the matrix related to the matrix of correlations between the variables of the two sets by the squared canonical correlation. It specifies the part of the mean variance in a set explained by a canonical variate when the other set is known. Thus, it tells how much redundant is a dataset if another set is given.

Table 6. Variances extracted and redundancies.

Specification	Environmental Conditions		Organic Farming	
	Variance Extracted	Redundancy	Variance Extracted	Redundancy
First canonical variate	0.0897	0.0296	0.1086	0.0359
Second canonical variate	0.0634	0.0167	0.0548	0.0145

The most statistically important canonical variate extracts nearly 11% of the variance in the set of variables related to the development level of organic farming and almost 9% in the set related to environmental conditions. In turn, the second canonical variate extracts around 6.3% of the variance in the set of variables related to environmental conditions and 5.5% in the set related to the development level of organic farming. The set of variables relating to environmental conditions makes it possible to explain 3.6% and 1.4%, respectively, of variance in the set of variables related to the development of organic farming. Conversely, the set of primary data relating to the development level of Polish organic farming can be used in explaining only 3.0% and 1.7%, respectively, in the second set under consideration, based on the two first statistically significant canonical variates. Therefore, already the second canonical variate has only a small contribution to explaining the variation.

Also, the study calculated the total redundancy which explains the average percentage of variation explained in one set of variables when another set is given (based on all canonical variates). It follows from these calculations that when the variables relating to environmental conditions are known, they can explain only 9.62% of the variance in the set of variables used in describing the development level of Polish organic farming. This suggests a weak dependency between the variables in both groups. In order to obtain better results, it would be worthwhile to carry out a study in the future with another set and number of variables and employ a procedure to weigh them.

The analysis also included drawing the dispersion graphs for the two statistically significant canonical roots (Figure 3). They show the relationship between the values of newly created variables relating to environmental conditions (the abscissa) and the development level of organic farming (the ordinate).

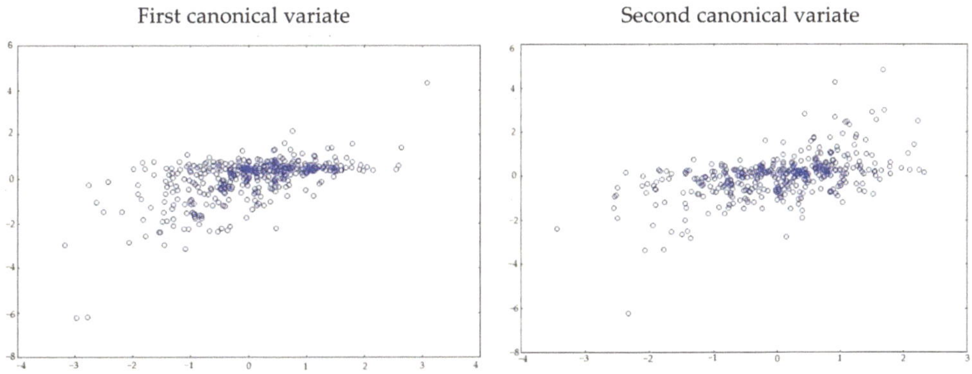

Figure 3. Dispersion graph for statistically significant canonical variates.

As regards the first statistically significant canonical variate, there is not much dispersion of points representing the objects (Polish districts) covered by the analysis. Most points are positioned along a straight line (with a slight positive slope). It can be assumed that the generated pairs of canonical variates convey a small part of information about the correlation between the two input datasets covered by the study. An increase in the value in the causative group (related to environmental variables) entails some overall growth in the group of effects (related to the development level of organic farming). As illustrated in the graph above, the relation is linear. A relatively strong concentration of points (representing the districts) in the graph could suggest the input variables share a similar structure. In the dispersion graph for the second canonical variate, the points representing the objects covered by the analysis are also positioned along a positively sloped (though near horizontal) line but are more dispersed with respect to it. The above provides grounds for concluding that the second pair of canonical variates conveys even less information on co-variability between the two variables considered.

5. Discussion

As empirically proven, organic products have a smaller content of pesticide and amino acid residues than food originating from other farming systems. However, findings from some research do not always reveal a smaller content of heavy metals (e.g., cadmium in beans, carrots, and apples) whose presence in plants may result from the condition of the environment where organic production takes place [55]. Environmental pollution and agricultural intensification (including the use of artificial fertilizers and plant protection products) are among the key threats to food quality. Indeed, some of the compounds found in foodstuffs are harmful to human body [56]. While the quality of an organic product is undeniably conditioned by a number of factors, particular attention should primarily be paid to production methods and to having a clean environment for cultivation and husbandry activities [57].

It follows from the analysis carried out at the overall European Union level that the relationship between the condition of the natural environment (quantified using the authors' own synthetic metric) and the number of organic producers and the share of organic farmland in total area of agricultural land is moderate, if not weaker. It is even weaker at the level of Polish smaller local government units, although Polish agriculture demonstrates some undisputable advantages (as mentioned in the theoretical part) from the perspective of the development potential it offers to organic farming, especially in regions at low industrialization levels (which also demonstrate a relatively unpolluted environment).

Thus, such a weak dependency discovered based on data gathered in Poland comes as somewhat of a surprise. Indeed, it follows from a study by J. Groszyk [18] based on 2011–2020 data that Polish organic farming thrives best in areas affected by difficult natural conditions where land-related limitations make conventional farming economically unviable. In turn, according to a study by J. Jętkowska [58], Polish NUTS-2 regions, which are leaders in the share of certified organic agricultural land, demonstrate a relatively low share of land under crops intended for organic food production. Hence, it cannot be asserted that the regions with the highest share of organic farmland may be used as an example of a good organic farming practice in the sense of sustainable development.

In the context of these considerations, it is also worth mentioning the findings from a study by W. Łuczka and S. Kalinowski [59], which suggest that when making their decision to go organic, Polish farmers take greater account of environmental aspects than reducing costs or improving their lifestyle. However, the key reason for doing so is the accessibility of financial support and the ability to sell their products at higher prices.

In other parts of the world, too, emphasis is placed on the importance of the natural environment as a determinant of organic farming development. As revealed by a study conducted in Iran, the farmers have a positive attitude towards the environment and are mostly willing to engage in organic farming. It was demonstrated that the acceptance of organic farming measures is strongly guided by the following: the farmers' intents related to going organic; their environmental identity; their responsibility for environmentally friendly behaviors; and their moral standards [60]. This is particularly important because, according to a study by M. Schleiffer and B. Speiser [61], absolutely all organic crops are affected by the potential risk of pesticide residue, and therefore organic farmers cannot adopt a zero-tolerance approach to that kind of content.

6. Conclusions

Although organic agricultural production keeps growing in popularity, the analysis carried out above suggests that its development differs between Union members. The countries largely differ in both the number of farms and the area of organic farmland. The factors behind this variation could include environmental conditions.

Implementing the goals of the European Green Deal requires taking radical measures in the agricultural sector which is carrying the burden of environmental (and climate) change, on the one hand, but has a strong impact on it, on the other. In a way, organic farming development is forced by the European Union's Common Agricultural Policy, and by the consumers becoming more and more aware of environmental protection issues and having the need to access healthy organic products. However, in countries such as Poland, reaching the target level of 25% of farmland under organic crops by 2030 requires the introduction of additional financial support instruments (e.g., a refund of certification costs for the producers or compensation for environmental protection expenditure).

Organic farming development provides greater opportunities for the rational use of natural resources and for offering high-quality food to consumers. That kind of farming stands out for being environmentally safe and is still highly valued by society (although the question remains whether it will continue to be the case after the farmer strikes held across Europe). On the one hand, organic farming is a response to the intensification of conventional farming, the deteriorating quality of food produced, the excessive use of mineral fertilizers and pesticides, and the related environmental pollution. On the other, it is significantly conditioned by environmental factors. Hence, in other words, the development of organic farming determines and (as a general rule) is determined by the natural environment.

Whether organic or conventional, farming faces the threat of pollutant emissions, including those generated by itself but mostly coming from non-agricultural sources, primarily from industrial and transportation sectors. In extreme cases, this may even involve the need to restrict agricultural production activities. The condition of the natural environment has an impact on the quality of food raw materials which may be the origin of different kinds of both biological and chemical threats. Guaranteeing that food is safe for human health takes on particular importance with respect to food made with organic methods, which should not use chemical protection in the production of raw materials in order to reduce the risk of biological threats.

According to empirical analyses carried out at the overall European Union level, a positive and statistically significant correlation (which can be viewed as moderate) exists between the condition of the natural environment, on the one side, and the number of organic producers and the share of organic farmland in total area of agricultural land, on the other. In turn, when restricting the study to the Polish territory only (note that Poland, as the only Union country, recorded a drop in the number of organic producers and in the share of organic farmland in total area of agricultural land between 2013 and 2021), there is a positive yet statistically insignificant correlation between the authors' synthetic metric of organic farming development and the condition of the environment (measured with TOPSIS metrics). As shown by the canonical analysis, when the variables relating to environmental conditions are known, they can explain only 9.62% of the variance in the set of variables used in describing the development level of Polish organic farming. Thus, the development level of organic development is driven by non-environmental conditions. The application objective of the article was to popularize canonical analysis. It should be emphasized again that the results of the canonical analysis are sensitive to outliers. Furthermore, obtaining reliable results requires a relatively large sample. It is particularly difficult to ensure the normality of the variables.

The results of this study may help to indirectly justify the need for implementing environmentally-oriented reforms at both the local and central levels, which seems to be necessary in order to reduce the negative consequences of environmental pollution and progressing global warming. Also worth considering could be a system for the environmental certification for regions, which would include adding the environmental score to the labels of local products. This is especially important as today's main environmental challenges, which give rise to common concerns, include atmospheric air pollution, water pollution, and uncontrolled pollution with waste.

Author Contributions: Conceptualization, M.M., L.S. and A.S.; methodology, M.M.; software, M.M.; validation, M.M., L.S. and A.S.; formal analysis, M.M., L.S. and A.S.; investigation, M.M., L.S. and A.S.; resources, M.M.; data curation, M.M.; writing—original draft preparation, M.M., L.S. and A.S.; writing—review and editing, M.M., L.S. and A.S.; visualization, M.M.; supervision, L.S. and A.S.; project administration, L.S. and A.S.; funding acquisition, M.M. All authors have read and agreed to the published version of the manuscript.

Funding: This research received no external funding.

Institutional Review Board Statement: Not applicable.

Data Availability Statement: Data available in a publicly accessible repository that does not issue DOIs. Publicly available datasets were analyzed in this study. This data can be found here: (Statistics Poland, Local Data Bank: https://bdl.stat.gov.pl/BDL/start (accessed on 14 July 2024); EUROSTAT: https://ec.europa.eu/eurostat/web/main/data/database (accessed on 14 July 2024); Database of the Inspectorate of Marketable Quality of Agri-Food Products: https://www.gov.pl/web/ijhars/rolnictwo-ekologiczne) (accessed on 14 July 2024).

Conflicts of Interest: The authors declare no conflicts of interest.

References

1. Von Weizsäcker, E.U.; Wijkman, A. *Come On! Capitalism, Short-Termism, Population and the Destruction of the Planet—A Report to the Club of Rome*; Springer: New York, NY, USA, 2018. [CrossRef]
2. Kristiansen, P. Overview of organic agriculture. In *Organic Agriculture: A Global Perspective*; Kristiansen, P., Taji, H., Reganold, J., Eds.; CSIRO Publishing: Clayton, Australia, 2006; pp. 1–23.

3. Vogt, G. The origins of organic farming. In *Organic Farming—An International History*; Lockeretz, W., Ed.; CABI: Oxfordshire, UK, 2007; pp. 9–29. [CrossRef]
4. Kuepper, G. *A Brief Overview of the History and Philosophy of Organic Agriculture*; Kerr Center for Sustainable Agriculture: Poteau, OK, USA, 2010.
5. Tomaš-Simin, M.; Glavaš-Trbić, D. Historical Development of Organic Production. *Econ. Agric.* **2016**, *3*, 1083–1098. [CrossRef]
6. European Commission. *AGENDA 2000 for a Stronger and Wider Union*; Office for Official Publications of the European Communities: Luxembourg, 1997.
7. European Commission. Communication from the Commission to the European Parliament, The European Council, The Council, The European Economic and Social Committee and the Committee of the Regions. The European Green Deal. COM(2019) 640 Final. 2019. Available online: https://eur-lex.europa.eu/resource.html?uri=cellar:b828d165-1c22-11ea-8c1f-01aa75ed71a1.0002.02/DOC_1&format=PDF (accessed on 9 September 2024).
8. Czudec, A.; Kata, R.; Miś, T. *Effects of the European Union's Agricultural Policy at the Regional Level*; Bogucki Wydawnictwo Naukowe: Poznań, Poland, 2017.
9. European Commission. Communication from the Commission to The European Parliament, The Council, The European Economic and Social Committee and the Committee of the Regions on an Action Plan for the Development of Organic Production. COM(2021) 141 Final. 2021. Available online: https://eur-lex.europa.eu/resource.html?uri=cellar:13dc912c-a1a5-11eb-b85c-01aa75ed71a1.0003.02/DOC_1&format=PDF (accessed on 9 September 2024).
10. Meemken, E.; Qaim, M. Organic Agriculture, Food Security, and the Environment. *Annu. Rev. Resour. Econ.* **2018**, *10*, 39–63. [CrossRef]
11. Azarbad, H. Conventional vs. Organic Agriculture-Which One Promotes Better Yields and Microbial Resilience in Rapidly Changing Climates? *Front. Microbiol.* **2022**, *13*, 903500. [CrossRef]
12. Gamage, A.; Gangahagedara, R.; Gamage, J.; Jayasinghe, N.; Kodikara, N.; Suraweera, P.; Merah, O. Role of organic farming for achieving sustainability in agriculture. *Farming Syst.* **2023**, *1*, 100005. [CrossRef]
13. Wilkin, J. Agriculture vs. climate. *Probl. Agric. Econ.* **2020**, *4*, 180–186.
14. Runowski, H. *Outcomes of Research on Socially Sustainable Farming. 2011–2014 Multiannual Program*; Zegar, J., Ed.; IERiGŻ-PIB: Warszawa, Poland, 2012; pp. 38–78.
15. Lorenz, K.; Lal, R. Environmental Impact of Organic Agriculture. *Adv. Agron.* **2016**, *139*, 99–152.
16. Łukasiński, W. Quality management of an organic product. *Food Sci. Technol. Qual.* **2008**, *1*, 146–153.
17. Łuczka., W. The State of the Art in Ecological Agriculture Research in Poland. *Res. Pap. Wrocław Univ. Econ.* **2017**, *453*, 64–76.
18. Groszyk, J. Organic farming in Poland in the context of Union strategies ch. *BAS* **2022**, *4*, 1–4.
19. Stolze, M.; Sanders, J.; Kasperczyk, N.; Madsen, G.; Meredith, S. *CAP 2014–2020: Organic Farming and the Prospects for Stimulating Public Goods*; IFOAM EU: Brussels, Belgium, 2016.
20. Miecznikowska-Jerzak, J. The status and prospects of organic farming in Poland—Assessment of challenges and opportunities for the implementation of the European Green Deal for agriculture. *Yearb. Eur. Integr.* **2022**, *16*, 265–283.
21. Kallas, Z.; Serra, T.; Gil, J.M. Farmer's objectives as determinant factors of organic farming adoption. *Res. Agric. Appl. Econ.* **2009**, *1*, 1–19. [CrossRef]
22. Siepmann, L.; Nicholas, K.A. German Winegrowers' Motives and Barriers to Convert to Organic Farming. *Sustainability* **2018**, *10*, 4215. [CrossRef]
23. Buys, P.; Chomitz, K.M.; De Luca, G.D.; Thomas, T.S.; Wertz-Kanounnikoff, S. *At Loggerheads Agricultural Expansion, Poverty Reduction, and Environment in the Tropical Forests*; A World Bank Policy Research Report; World Bank Group: Washington, DC, USA, 2006.
24. Lapola, D.; Martinelli, L.; Peres, C.; Ometto, J.P.H.B.; Ferreira, M.E.; Nobre, C.A.; Aguiar, A.P.D.; Bustamante, M.M.C.; Cardoso, M.F.; Costa, M.H.; et al. Pervasive transition of the Brazilian land-use system. *Nat. Clim. Change* **2014**, *4*, 27–35. [CrossRef]
25. Kociszewski, K. Barriers and Factors Favorable for Functioning of Organic Farms in the Light of Nationwide Questionnaire Survey. *Ann. Pol. Assoc. Agric. Agribus. Econ.* **2014**, *16*, 129–134.
26. Siedlecka, A. Conditions and Prospects for The Development of Organic Farms in Natural Valuable Areas of The Lubelskie Province. *Ann. Pol. Assoc. Agric. Agribus. Econ.* **2015**, *17*, 240–245.
27. Kozłowska-Burdziak, M.; Gardocka-Jałowiec, A. Conditions affecting development of ecological agriculture in Podlasie Voivodeship. *Issues Agric. Advis. Serv.* **2018**, *1*, 55–66.
28. Supreme Chamber of Control. *Supporting the Development of Organic Farming*; Department of Agriculture and Rural Development: Warsaw, Poland, 2018.
29. Łuczka, W.; Kalinowski, S. Socioeconomic Reasons for Discontinuing Organic Farming: A Polish Case Study. *Acta Sci. Polonorum. Oeconomia* **2023**, *22*, 27–46. [CrossRef]
30. Ashari, N.F.N.; Sharifuddin, J.; Abidin, Z.A. Factors Determining Organic Farming Adoption: International Research Results and Lessons Learned for Indonesia. *Forum Penelit. Agro Ekon.* **2017**, *35*, 45–58. [CrossRef]
31. Cukur, T.; Kizilaslan, N.; Kizilaslan, H. Analysis Of The Factors Affecting The Adoption of Organic Farming in Turkey: The Case of Samsun province. *Appl. Ecol. Environ. Res.* **2020**, *17*, 14001–14008. [CrossRef]
32. Karipidis, P.; Karypidou, S. Factors that Impact Farmers' Organic Conversion Decisions. *Sustainability* **2021**, *13*, 4715. [CrossRef]

33. Kujala, S.; Hakala, O.; Viitaharju, L. Factors affecting the regional distribution of organic farming. *J. Rural Stud.* **2022**, *92*, 226–236. [CrossRef]
34. Riar, A.; Goldmann, E.; Bautze, D.; Rüegg, J.; Bhullar, G.B.; Adamtey, N.; Schneider, M.; Huber, B.; Armengot, L. Farm gate profitability of organic and conventional farming systems in the tropics. *Int. J. Agric. Sustain.* **2024**, *22*, 2318933. [CrossRef]
35. Hwang, C.L.; Yoon, K. *Multiple Attribute Decision Making: Methods and Applications*; Springer: Berlin/Heidelberg, Germany, 1981.
36. Krohling, R.A.; Pacheco, A.H.G. A-TOPSIS—An approach Based on TOPSIS for Ranking Evolutionary Algorithms. *Procedia Comput. Sci.* **2015**, *55*, 308–317. [CrossRef]
37. Madanchian, M.; Taherdoost, H. A comprehensive guide to the TOPSIS method for multi-criteria decision making. *Sustain. Soc. Dev.* **2023**, *1*, 2220. [CrossRef]
38. Młodak, A. An application of a complex measure to model–based imputation in business statistics. *Stat. Transit. New Ser.* **2021**, *22*, 1–28. [CrossRef]
39. Timm, N.H. *Applied Multivariate Analysis. Springer Texts in Statistics*; Springer: New York, NY, USA, 2002.
40. Hardoon, D.R.; Szedmak, S.; Shawe-Taylor, J. Canonical Correlation Analysis. In *An Overview with Application to Learning Methods*; University of London: London, UK, 2003.
41. Legendre, P.; Legendre, L. *Developments in Environmental Modelling*; Elsevier: Amsterdam, The Netherlands, 2012.
42. Bilenko, N.Y.; Gallant, J.L. Regularized Kernel Canonical Correlation Analysis in Python and Its Applications to Neuroimaging. *Front. Neuroinform.* **2016**, *10*, 49. [CrossRef]
43. Abdi, H.; Guillemot, V.; Eslami, A.; Beaton, D. Canonical Correlation Analysis. In *Encyclopedia of Social Network Analysis and Mining*; Alhajj, R., Rokne, J., Eds.; Springer: New York, NY, USA, 2018. [CrossRef]
44. Panek, T.; Zwierzchowski, J. *Statistical Methods for Multidimensional Benchmarking: Theory and Use Cases*; SGH Publishing House: Warsaw, Poland, 2013.
45. Cliff, O.M.; Novelli, L.; Fulcher, B.D.; Shine, J.M.; Lizier, J.T. Assessing the Significance of Directed and Multivariate Measures of Linear Dependence Between Time Series. *Phys. Rev. Res.* **2021**, *3*, 013145. [CrossRef]
46. Leys, C.; Ley, C.; Klein, O.; Bernard, P.; Licata, L. Detecting outliers: Do not use standard deviation around the mean, use absolute deviation around the median. *J. Exp. Soc. Psychol.* **2013**, *49*, 764–766. [CrossRef]
47. Box, G.E.P.; Cox, D.R. An analysis of transformations. *J. R. Stat. Soc. Ser. B* **1964**, *26*, 211–252. [CrossRef]
48. Sadowski, A.; Wojcieszak-Zbierska, M.; Zmyślona, J. Economic Situation of Organic Farms in Poland on the Background of the European Union. *Probl. Agric. Econ.* **2021**, *367*, 101–118. [CrossRef]
49. Eurostat Website. Available online: https://ec.europa.eu/eurostat/web/main/data/database (accessed on 16 July 2024).
50. Willer, H.; Schlatter, B.; Trávníček, J. *The World of Organic Agriculture. Statistics and Emerging Trends 2023*; Research Institute of Organic Agriculture FiBL, Frick, and IFOAM—Organics International: Bonn, Germany, 2023.
51. Willer, H.; Schlatter, B.; Meier, C.; Trávníček, J. *The World of Organic Agriculture. Statistics and Emerging Trends 2021*; Research Institute of Organic Agriculture FiBL, Frick, and IFOAM—Organics International: Bonn, Germany, 2021.
52. Heinze, S.; Vogel, A. Reversion from Organic to Conventional Agriculture in Germany: An Event History Analysis. *Ger. J. Agric. Econ.* **2017**, *66*, 13–25. [CrossRef]
53. Central Statistical Office Website. Available online: https://bdl.stat.gov.pl/BDL/start (accessed on 14 July 2024).
54. Database of the Inspectorate of Marketable Quality of Agri-Food Products (IJHARS). Available online: https://www.gov.pl/web/ijhars/rolnictwo-ekologiczne (accessed on 14 July 2024).
55. Żurek, N.; Bilek, M. Health safety of organic food in the light of research and inspection activities. *Bromatol. I Chem. Toksykol.* **2018**, *51*, 1–7.
56. Staniek, S. Characteristics of food produced in organic farming conditions. *Pol. J. Agron.* **2014**, *19*, 25–35.
57. Kaźmierczak-Piwko, L. Ecological Risk as a Component of the Credit Risk of Financial Liabilities of Manufacturing Companies. *Syst. Support. Prod. Eng.* **2017**, *6*, 61–68.
58. Jętkowska, J. Organic farming in Poland. The share of organic agricultural land in the context of the European Green Deal. *Issues Agric. Advis. Serv.* **2022**, *107*, 5–21.
59. Łuczka, W.; Kalinowski, S. Barriers to the Development of Organic Farming: A Polish Case Study. *Agriculture* **2020**, *10*, 536. [CrossRef]
60. Fatemi, M.; Rezaei-Moghaddam, K. Sociological factors influencing the performance of organic activities in Iran. *Life Sci. Soc. Policy* **2020**, *16*, 1–16. [CrossRef]
61. Schleiffer, M.; Speiser, B. Presence of pesticides in the environment, transition into organic food, and implications for quality assurance along the European organic food chain—A review. *Environ. Pollut.* **2022**, *313*, 120116. [CrossRef]

agriculture

MDPI

Article

The Role of Red Clover and Manure Fertilization in the Formation of Crop Yield of Selected Cereals

Irena Suwara [1,*], Katarzyna Pawlak-Zaręba [1], Dariusz Gozdowski [2] and Renata Leszczyńska [1]

1 Department of Agronomy, Institute of Agriculture, Warsaw University of Life Sciences, Nowoursynowska 159, 02-776 Warsaw, Poland; katarzynapawlak7@gmail.com (K.P.-Z.); renata_leszczynska@sggw.edu.pl (R.L.)

2 Department of Biometry, Institute of Agriculture, Warsaw University of Life Sciences, Nowoursynowska 159, 02-776 Warsaw, Poland; dariusz_gozdowski@sggw.edu.pl

* Correspondence: irena_suwara@sggw.edu.pl

Abstract: The use of legumes in rotation is beneficial and is of great importance in sustainable agricultural production in line with the assumptions of the European Green Deal. The aim of the presented research was to evaluate the cultivation of red clover as an undersown crop for spring barley and as a forecrop for winter wheat on the yield and quality of spring barley and winter wheat. To achieve this goal, two long-term static experiments set up in 1955 were used, in which diversified mineral and organic fertilization were used in two rotations: rotation without red clover (sugar beet–spring barley–winter rapeseed–winter wheat) and rotation with red clover (sugar beet–spring barley with undersown red clover–red clover–winter wheat). The obtained results indicate that the Norfolk rotation with red clover, as well as varied fertilization and years of research, influence the yield of plants. The highest grain yields of spring barley (5.7 t ha^{-1}) were ensured by mineral fertilization (NPK) and mineral fertilization in combination with manure ($\frac{1}{2}$NPK + $\frac{1}{2}$FM). However, the highest yields of winter wheat grain (6.4 t ha^{-1}) were recorded in the treatments with exclusive mineral fertilization (NPK), significantly lower yields in the treatments where mineral fertilizers were used in combination with manure (5.7 t ha^{-1}) ($\frac{1}{2}$NPK + $\frac{1}{2}$FM) and only manure (5.1 t ha^{-1}) (FM). The lowest yields of both cereals were found on soil that had not been fertilized since 1955 (0). The grain yield of spring barley was not significantly differentiated by the sowing method and was similar for spring barley grown with and without undersown red clover. Including legumes in the rotation had a positive effect on the yield of winter wheat. Fertilization had the greatest impact on the protein content in cereal grains. The use of mineral fertilization (NPK) and mineral fertilization in combination with manure ($\frac{1}{2}$NPK + $\frac{1}{2}$FM) ensured the highest protein content in the grain of spring barley and winter wheat. Mineral fertilization (NPK) increased the protein content in spring barley grain by 2.9 percentage points compared to the unfertilized treatment (0) and by 2.1 percentage points compared to exclusive manure fertilization (FM), and in winter wheat grain by 2.3 and 1.4 percentage points, respectively. The cultivation of red clover in the rotation also had a positive effect on the protein content in spring barley and winter wheat grains.

Keywords: crop yield; farmyard manure FM; mineral fertilization NPK; crop rotation; red clover

1. Introduction

Cereals like wheat (*Triticum aestivum* L.) and barley (*Hordeum vulgare* L.) are the major and most important crops in many countries. There are many attempts to increase wheat and barley productivity. It is widely known that a well-designed crop rotation improves soil structure, better utilization of nutrients by plants, reduces the occurrence of weeds, pests, and disease, and thus increases plant yields, including cereals [1–4]. For cereal plants such as winter wheat and spring barley, the forecrop is very important, the improper selection of which results in a significant reduction in yield [5,6]. Research by Suwara et al. [7] showed a beneficial effect of legumes on the yield of winter wheat. A significant share of cereal plants in the rotation leads to a reduction in the yield and deterioration of its quality, both in the case of spring barley and winter wheat [8–11]. The amount of yield obtained may be determined not only by rotation but also by fertilization. The yield of cereal plants largely depends on the amount of macro- and microelements accumulated in them [12] and the availability of nutrients such as nitrogen, potassium, and phosphorus [6,13,14].

The use of legumes and organic fertilization in rotation is beneficial and is of great importance in sustainable agricultural production in accordance with the assumptions of the European Green Deal [15,16]. Sustainable development is a way of meeting the needs of the current generation while not limiting the production potential of future generations [17]. According to Czyżewski et al. [18], the development of sustainable agriculture is one of the most important issues in modern agricultural economics. One of the main goals of sustainable agriculture is to reduce environmental pollution with chemical components from mineral fertilizers by adapting fertilization to the needs of plants and soil conditions. The essence of sustainable agriculture is not only the rational use of crop fertilization but concern for the protection of soil productivity [19,20]. Norfolk rotation and organic fertilization protect the soil against degradation because both fertilization and rotation have a positive effect on soil properties, which determine the course of a number of soil processes, including the supply of plants with water, air, and nutrients [21–24]. Using organic fertilizers instead of mineral fertilizers is an environmentally friendly practice that is very important in sustainable agricultural systems. The main advantage of organic fertilizers is that they are obtained from organic materials, i.e., plant remains, animal excrements, and food industry by-products. Organic fertilizers are cheap, improve soil structure and aeration, and increase porosity and the soil's ability to retain water. Additionally, manure is known to reduce the rate of evaporation, stimulate root development, and optimize plant growth. In summary, manure consistently provides nutrients to crops through a natural biological process [25–28]. Sustainable agriculture plays a decisive role in adapting to climate change as well as achieving sustainable development goals [29]. The availability of soil water for plants and the retention capacity of the soil in conditions of climate change, in addition to fertilization, are the basic elements determining plant yields.

The special role of legumes in the sustainable agriculture system results, among other things, from their ability to fix atmospheric nitrogen thanks to symbiosis with nodule bacteria. In practice, this could mean large savings resulting from limiting the use of nitrogen in mineral form [30–32]. Moreover, their cultivation has a positive effect on improving the soil structure and enriching it with large amounts of organic matter due to the huge amount of crop residue left behind [33,34]. Cereals are often grown with undersown cover crops, which are mainly small-seeded legumes, which are of great importance in achieving the goals of the Green Deal. The use of undersown cereal crops limits weed infestation, reduces the degree of disease infection, and eliminates the unfavorable effects resulting from the succession of cereal crops. Legumes are also an excellent forecrop for subsequent crops because they leave a large mass of post-harvest residues rich in nitrogen. Nitrogen stored in the roots of these plants accounts for over 25% of the total nitrogen taken up by legumes [35–37].

In order to obtain better yields and produce high-quality grains, it is recommended to use organic fertilizers in plant cultivation. Various organic fertilizers should be used combined with mineral fertilizers for the purpose of improving cereal productivity and achieving the optimal level of agricultural sustainability [38,39]. The effects of fertilization and crop rotation are best assessed based on long-term field experiments, which give a unique possibility to analyze changes in soils, plants, and ecosystems [40–43]. The aim of the presented research was to evaluate the cultivation of red clover as an undersowing for spring barley and a forecrop for winter wheat in two long-term static field experiments established in 1955 at the experimental field of the Warsaw University of Life Sciences in Chylice, central Poland, on the yield and quality of spring barley and winter wheat.

2. Materials and Methods

This paper presents a yield analysis based on the results obtained for spring barley from 2011, 2015, and 2019 and for winter wheat from 2009, 2013, and 2021. To achieve this goal, two long-term static experiments were used, in which diversified mineral and organic fertilization were used in two rotations: rotation without red clover (sugar beet–spring barley–winter rapeseed–winter wheat) and Norfolk rotation with red clover (sugar beet–spring barley with undersown red clover–red clover–winter wheat).

The basis of the research was two long-term static field experiments established in 1955 at the Agricultural Experimental Station of the SGGW Chylice in Jaktorów. They are located in Central Poland, in the Masovian Lowlands, approximately 40 km west of Warsaw, in a plain landscape, elevated approximately 105 m above sea level (52°06′ N, 20°33′ E). The experiments were carried out on leached black earth [44] (according to the World References Base for Soil Resources WRB–Endogleyic Phaeozems), which was formed from light boulder clay. The density of the solid phase of this soil is 2.62 $g \cdot cm^{-3}$, and the humus horizon has a thickness of 30–35 cm. This soil is characterized by medium humus content, slightly acidic reaction, and regulated water relations. Table 1 shows the characteristics of the topsoil properties after 40 years of experiments. Before the experiments were carried out, the arable layer was slightly acidic (pH 6.2–6.5) and contained 1.15% organic carbon, 44 mg kg^{-1} of available phosphorus (P), and 83 mg kg^{-1} of available potassium (K). In the two experiments, four fertilizer treatments were compared: mineral fertilization (NPK), farmyard manure (FM), mixed mineral and organic fertilization ($\frac{1}{2}$ NPK + $\frac{1}{2}$FM), and control without any fertilization (0). These four treatments were investigated in a randomized block-design trial with four replicates. Fertilizers were applied in two crop rotations: Fertilization of particular crops is presented in Tables 2 and 3. The following mineral fertilizers were used for fertilization: ammonium nitrate (34% N), granulated superphosphate (18–19% P_2O_5), and potassium chloride (60% K_2O). Composted cattle manure is plowed into the soil in autumn. Mineral fertilizers (NPK) are used before sowing crops. In the case of winter wheat, the first dose of nitrogen (30% of the full dose) was applied before sowing together with P and K fertilizers, and the second dose (70% of the full dose) was applied in the tillering phase. In spring barley, the first dose of nitrogen (30% of the full dose) was applied before sowing together with P and K fertilizers, and the second dose (70% of the full dose) was applied after plant emergence. A plow tillage system was used in the experiments (plowing depth of 20 cm). Plant protection products (pesticides) were applied according to the needs of the plants. Both cereals were harvested after reaching full maturity with grain moisture below 18% and most often took place in the first half of August.

Table 1. Characteristics of the arable layer of the black earth in Chylice–pH, organic carbon content, and soil abundance in available forms of nutrients depending on the fertilization system and crop rotation.

Treatment	pH in KCL	P [$mg \cdot kg^{-1}$]	K [$mg \cdot kg^{-1}$]	C org. [$g \cdot kg^{-1}$]
		Fertilization		
NPK	6.1	79.4	68.9	10.05
FM	6.4	74.6	126.2	12.39
$\frac{1}{2}$ NPK + $\frac{1}{2}$ FM	6.3	80.7	83.0	11.30
0	6.3	49.3	48.1	8.91
		Crop rotation		
Norfolk rotation	6.1	65.0	77.2	12.34
Rotation without legumes	6.4	77.2	86.3	8.75

Mineral fertilization (NPK), farmyard manure (FM), mixed mineral and organic fertilization ($\frac{1}{2}$ NPK + $\frac{1}{2}$FM), control without any fertilization (0).

Table 2. Diagram of fertilizer experiments in Norfolk rotation.

Crop	Treatment								
	NPK			FM		$\frac{1}{2}$ NPK + $\frac{1}{2}$ FM			O
	N	P	K	FM	FM	N	P	K	
	[kg·ha^{-1}]			[t·ha^{-1}]	[t·ha^{-1}]	[kg·ha^{-1}]			
Sugar beet	200	56.0	200.0	40	20	100	28.0	100.0	0
Spring barley with red clover	100	36.5	91.5	20	10	50	18.3	45.8	0
Red clover	0	36.5	91.5	0	0	0	18.3	45.8	0
Winter wheat	100	36.5	91.5	20	10	50	18.3	45.8	0

Mineral fertilization (NPK), farmyard manure (FM), mixed mineral and organic fertilization ($\frac{1}{2}$ NPK + $\frac{1}{2}$FM), control without any fertilization (0).

Table 3. Diagram of fertilizer experiments in crop rotation without legumes.

Crop	Treatment								
	NPK			FM		$\frac{1}{2}$ NPK + $\frac{1}{2}$ FM			O
	N	P	K	FM	FM	N	P	K	
	[kg·ha^{-1}]			[t·ha^{-1}]	[t·ha^{-1}]	[kg·ha^{-1}]			
Sugar beet	200	56.0	200.0	40	20	100	28.3	100.0	0
Spring barley	100	36.5	91.5	20	10	50	18.3	45.8	0
Winter rapeseed	100	36.5	91.5	20	10	50	18.3	45.8	0
Winter wheat	100	36.5	91.5	20	10	50	18.3	45.8	0

Mineral fertilization (NPK), farmyard manure (FM), mixed mineral and organic fertilization ($\frac{1}{2}$ NPK + $\frac{1}{2}$FM), control without any fertilization (0).

Meteorological data on temperature and precipitation in Chylice in the years 2009, 2011, 2013, 2015, 2019, and 2021 are summarized in Tables 4 and 5. In 2011, 2015, and 2019, when spring barley was cultivated, the most favorable rainfall and thermal conditions were recorded in 2019. It was a warm year with good rainfall distribution during the spring barley growing season. In 2011, excessive rainfall in July had a negative impact on the ripening and harvesting of this cereal. However, 2015 was relatively dry, with a cold spring and very low rainfall recorded in June. In the years of winter wheat cultivation (2009, 2013, and 2021), the best moisture conditions for the growth and development of this plant occurred in 2021. In 2009 and 2013, unfavorable moisture conditions were found in the spring growing season of winter wheat due to excessive rainfall in May (2013) and June (2009 and 2013). Moreover, in April 2009, an extreme drought was recorded.

Table 4. Sum of precipitation in Chylice in 2009–2021 compared with the long-term average (1921–2020) data [mm].

Year	Sum	Month											
		I	II	III	IV	V	VI	VII	VIII	IX	X	XI	XII
2009	713.6	37.4	47.2	60.6	14.1	79.4	114.5	90.7	78.1	17.4	82.8	53.1	42.8
2011	693.5	37.0	27.1	15.1	78.2	48.1	57.4	251.4	118.7	6.9	14.2	1.6	37.8
2013	825.8	71.1	39.9	54.7	49.8	126.9	211.6	23.2	60.5	82.2	33.0	45.5	27.4
2015	421.3	43.5	12.1	26.0	45.8	51.2	16.1	64.0	6.4	37.2	44.5	56.2	18.5
2019	516.9	36.0	34.7	34.2	31.9	47.5	24.1	64.3	69.5	79.1	21.6	13.6	40.4
2021	685.4	88.9	88.4	17.6	59.5	58.2	46.4	135.1	179.2	29.6	8.9	47.2	33.4
Averaged monthly sums for 1955–2001		25.5	28.9	31.4	44.6	56.6	76.6	87.4	56.9	58.0	37.8	40.3	35.6

Table 5. Mean temperature in 2009–2021 compared with the long-term average (1921–2020) data [°C].

Year	Average	Month											
		I	II	III	IV	V	VI	VII	VIII	IX	X	XI	XII
2009	7.4	−3.9	−1.6	1.7	9.6	11.9	15.4	17.6	16.9	13.3	5.3	4.5	−2.3
2011	8.6	−1.0	−5.5	2.3	9.0	14.9	18.5	17.7	19.2	14.6	8.8	2.2	2.2
2013	8.1	−3.9	−1.2	−2.6	7.2	17.7	17.4	17.9	18.1	10.8	9.2	4.8	1.7
2015	9.3	0.5	0.5	4.4	7.5	13.0	16.3	18.8	21.1	14.1	6.4	4.6	4.1
2019	10.5	−1.6	3.2	6.2	9.8	13.2	22.0	19.0	20.1	14.0	10.4	6.0	3.1
2021	8.3	−1.5	−2.5	3.0	6.4	12.1	19.4	20.9	16.6	13.1	8.7	4.7	−1.3
Averaged monthly means 1955–2001		−1.8	−0.4	2.3	8.5	14.1	17.3	18.7	18.3	13.2	9.2	3.0	−1.1

The yield of cereal plants was determined by collecting winter wheat and spring barley plants after full grain maturity (BBCH 89) from each experimental plot with an area of 50 m^2 and converting them into grain yield per 1 ha at 14% humidity. Then, the quality parameters of winter wheat and spring barley grain were assessed in the laboratory using the Infratec 1241 grain analyzer from FOSS Analytics (Hilleroed, Denmark). It is a whole-grain analyzer that uses the absorption of near-infrared radiation to simultaneously determine various grain quality parameters at the same time. Measurements are performed in the wavelength range 570–1055 nm. Grain parameters were determined using this analyzer: protein content [%], wet gluten efficiency [%], starch content [%], and Zeleny sedimentation index [cm^3] [https://www.fossanalytics.com/en/products/infratec (accessed on 15 October 2024)].

In this study, the results of yield and quality characteristics are given as averages over the years for spring barley from 2011, 2015, and 2019 and for winter wheat from 2009, 2013, and 2021. For the tested parameters, averages over the years of research were calculated to compare the impact of the studied factors, i.e., fertilization and rotation. For three years, a three-way analysis of variance (ANOVA) was performed where the factors were crop rotation, fertilization, and year. Comparisons of means were performed using the Tukey procedure, and NIR values were calculated at a significance level of 0.05. On the basis of these analyses, homogeneous groups of means were distinguished, i.e., groups of means that did not differ significantly statistically were marked with the same letter of the alphabet. p-values were presented for selected traits for evaluation of the main effects of the studied factors as well their interactions, including interaction with years. In all analyses, the significance level was set at 0.05. Analyses were performed in Statistica 13 (TIBCO Software Inc., Palo Alto, CA, USA) [45,46].

3. Results and Discussion

The obtained results indicate that the Norfolk rotation with red clover, as well as varied fertilization and years of research, influence the yield of plants. It was found that the yield of spring barley is mainly determined by the fertilization system (Figure 1). The use of NPK and $\frac{1}{2}$NPK + $\frac{1}{2}$FM clearly stimulated the productive tillering of spring barley in both experiments. The number of spring barley ears per square meter was on average 556 in the soil fertilized only with mineral fertilizers (NPK), 587 with mineral fertilizers including manure ($\frac{1}{2}$NPK + $\frac{1}{2}$FM), and 528 in the plots fertilized only with manure (FM). The lowest number of ears was found in the unfertilized plot (336). In all fertilized treatments (NPK, FM, and $\frac{1}{2}$NPK + $\frac{1}{2}$FM), the highest thousand-grain weight (49.6–51.1 g) was recorded compared to the treatment that had not been fertilized since 1955 (45.9 g). As a result, the highest yields of spring barley grain were ensured by mineral fertilization (NPK) and mineral fertilization combined with manure ($\frac{1}{2}$NPK + $\frac{1}{2}$FM), which is in agreement with reports in the literature [47–50].

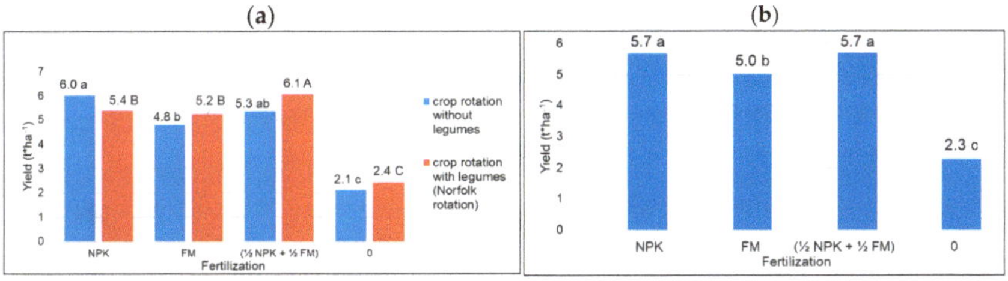

Figure 1. Spring barley grain yield depending on fertilization in two rotations (**a**) and only fertilization (**b**), means of years 2011, 2015, and 2019. Individual letters indicate homogeneous groups of means, mean values marked with the same letters do not differ significantly at α = 0.05, different letters indicate significant differences at α = 0.05. In subfigure (**a**), lowercase letters refer to crop rotation without legumes; capital letters refer to crop rotation with legumes (Norfolk rotation) Table.

Significantly lower yields were found in the area fertilized only with manure (FM), and the smallest were found in soil unfertilized since 1955 (0) (Figure 1).

The grain yield of spring barley was not significantly altered by the sowing method and was similar for spring barley grown with and without undersown red clover. Spring barley yielded at a similar level, regardless of the use of undersowing or not (Figure 1). Also, Alaru et al. [16] found that red clover as an undersow in spring barley had no significantly positive effect on the grain yield and protein content of barley. In turn, Wanic et al. [51] noted that the number of barley ears at the end of the vegetation period in pure sowing was significantly higher than with underseeds, and, as a result, spring barley grown with underseeds yielded worse than in pure sowing. In the study by Andruszczak et al. [52], when growing spring barley in monoculture, undersowing of red clover promoted spring barley yield by 24.0% compared to barley in pure sowing.

The yield of winter wheat depended on both fertilization and rotation. The use of NPK clearly stimulated the productive tillering of winter wheat in both experiments, and in this treatment, the highest grain yields of winter wheat were recorded (Figure 2). In the plots fertilized with mineral fertilizers together with manure ($\frac{1}{2}$NPK + $\frac{1}{2}$FM), winter wheat yielded on average about 10% lower, and in the plots fertilized only with manure (FM), the yield was over 20% lower.

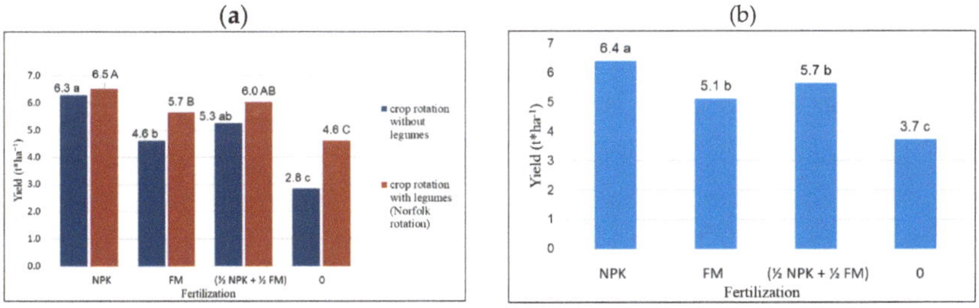

Figure 2. Winter wheat grain yield depending on fertilization in two rotations (**a**) and only fertilization (**b**), means of years 2011, 2015, and 2019. Individual letters indicate homogeneous groups of means, mean values marked with the same letters do not differ significantly at α = 0.05, different letters indicate significant differences at α = 0.05. In subfigure (**a**), lowercase letters refer to crop rotation without legumes; capital letters refer to crop rotation with legumes (Norfolk rotation).

Some researchers received different results [53,54]. Jiang et al. [53] found the highest yields of wheat with organic fertilizers combined with NPK and almost 1 t ha^{-1} higher yields when compared to NPK without organic compounds. Also Blecharczyk et al. [55], Ailincăi et al. [56] and Barzegar et al. [57] obtained higher grain yields with NPK incorporated with farmyard manure in comparison with NPK alone.

Yields were the lowest in unfertilized soil (0), significantly so in relation to fertilized treatments. It should be emphasized that winter wheat in the Norfolk rotation yielded relatively well on plots that had not been fertilized since 1955. Yields of winter wheat on unfertilized plots with red clover as a forecrop were, on average, about 50–60% higher compared to wheat grown after winter rapeseed (Figure 2a). The grain yield of winter wheat was significantly differentiated by forecrop. Including legumes in the rotation had a positive effect on the yield of winter wheat. Winter wheat yields in the rotation with red clover were 20% higher than in the rotation without legume (on average 5.7 t ha^{-1} vs. 4.8 t ha^{-1}, respectively) (Figure 3). Winter wheat grown after red clover produced a greater number of ears per square meter (504) than wheat after winter rapeseed (456) Including legumes in the rotation had a positive effect on the yield of winter wheat. This is evidenced by significantly higher grain yields of winter wheat grown after red clover (Figure 3). The beneficial effect of legumes on wheat grain yields was also noted by Berzsenyi et al. [58], Norwood [59], Blecharczyk et al. [60], Buczek et al. [61], Smagacz and Kuś [62], Amato et al. [63], and Małecka-Jankowiak et al. [64].

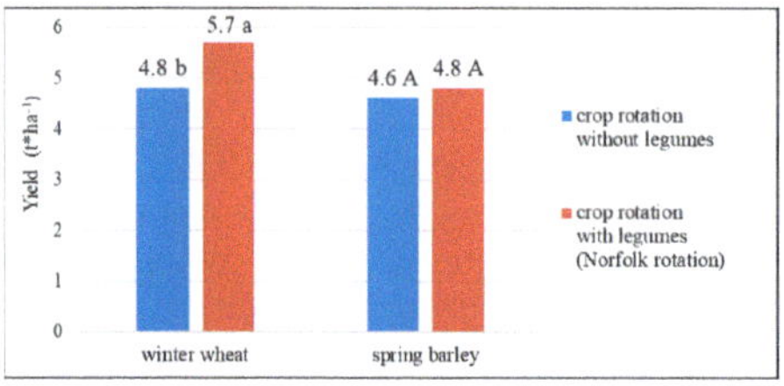

Figure 3. Winter wheat grain yield (means of years 2009, 2013, and 2021) and spring barley (means of years 2011, 2015, and 2019) depending on the crop rotation. Individual letters indicate homogeneous groups of means, mean values marked with the same letters do not differ significantly at $\alpha = 0.05$, different letters indicate significant differences at $\alpha = 0.05$; lowercase letters refer to winter wheat; capital letters refer to spring barley.

In order to determine the quality of wheat grain, the content of protein, gluten, starch, and the Zeleny index were investigated. The quality characteristics of wheat are important to consumers, growers, millers, and bakers. The quality of wheat is determined by its protein and gluten content. The quality of flour and dough is influenced by starch content, gluten, falling number, and dough rheology. A high protein content in wheat grain improves the structure and volume of the bread, while gluten has an impact on the stability of the dough during baking [65]. For spring barley grain, the main uses are in the brewing and feed industries. Because barley does not contain gluten, it is used to a lesser extent in food production, e.g., as an admixture in bread making. Due to its high starch and fiber content and moderate protein content, it is popular for feeding ruminant animals. For the brewing industry, one of the most relevant factors is protein content in grain [66,67].

The results presented in Figures 4–9 indicate that fertilization significantly affects the technological parameters of spring barley grain and winter wheat grain. The quality of spring barley grain depended mainly on fertilization and the presence of underseed red clover. Fertilization had the greatest impact on the protein content in grain (Figure 4). The use of mineral fertilization and mineral fertilization in combination with manure ensured the highest protein content in spring barley grain. The lowest protein level was recorded in grain from unfertilized treatments and those fertilized only with manure. In turn, originating grain contained the most starch from unfertilized treatments and fertilized only with manure, and the least in grain from mineral fertilized treatments (Figure 5).

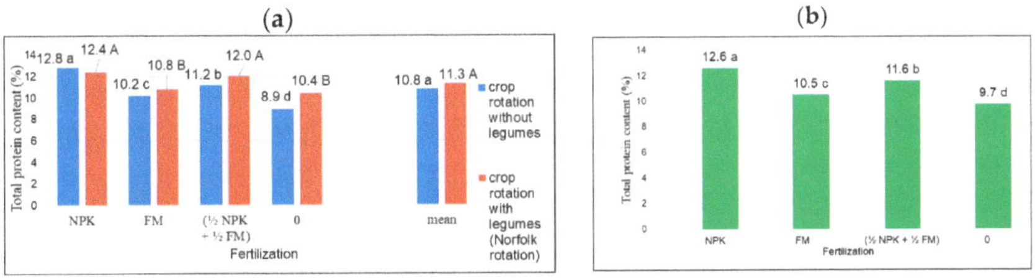

Figure 4. Protein content in spring barley grain depending on fertilization in two rotations (**a**) and only fertilization (**b**), means of years 2011, 2015, and 2019. Individual letters indicate homogeneous groups of means, mean values marked with the same letters do not differ significantly at $\alpha = 0.05$, different letters indicate significant differences at $\alpha = 0.05$. In subfigure (**a**), lowercase letters refer to crop rotation without legumes; capital letters refer to crop rotation with legumes (Norfolk rotation).

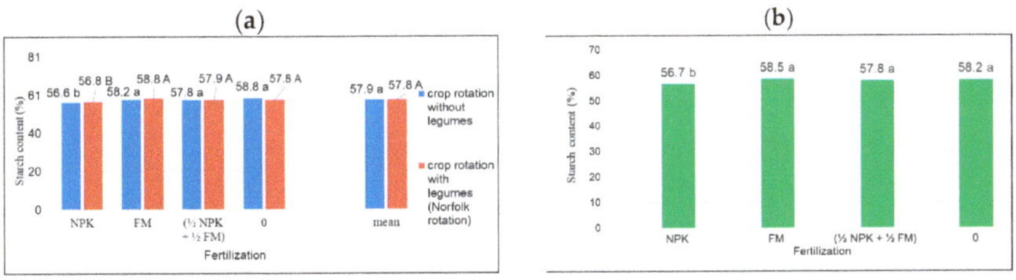

Figure 5. Starch content in spring barley grain depending on fertilization in two rotations (**a**) and only fertilization (**b**), means of years 2011, 2015, and 2019. Individual letters indicate homogeneous groups of means, mean values marked with the same letters do not differ significantly at $\alpha = 0.05$, different letters indicate significant differences at $\alpha = 0.05$. In subfigure (**a**), lowercase letters refer to crop rotation without legumes; capital letters refer to crop rotation with legumes (Norfolk rotation).

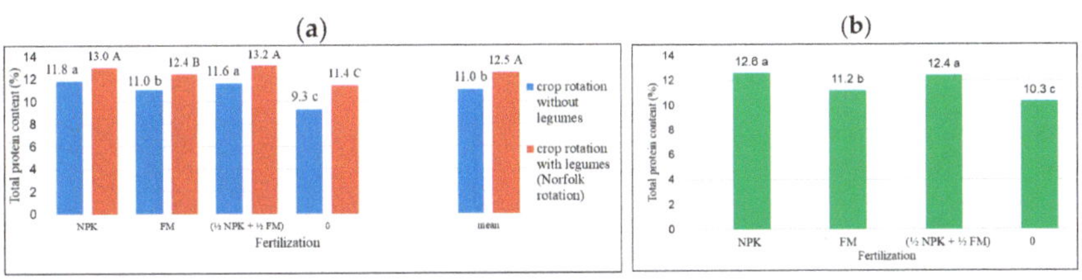

Figure 6. Protein content in winter wheat grain depending on fertilization in two rotations (**a**) and only fertilization (**b**), means of years 2011, 2015, and 2019. Individual letters indicate homogeneous groups of means, mean values marked with the same letters do not differ significantly at $\alpha = 0.05$, different letters indicate significant differences at $\alpha = 0.05$. In subfigure (**a**), lowercase letters refer to crop rotation without legumes; capital letters refer to crop rotation with legumes (Norfolk rotation).

The results in Figures 6–8 indicate that fertilization significantly affects the technological parameters of winter wheat grain, primarily the content of total protein and wet gluten. The highest protein and wet gluten content, as well as the highest Zeleny sedimentation index, were found in winter wheat grain fertilized with minerals (NPK and $\frac{1}{2}$NPK + $\frac{1}{2}$FM) and then fertilized only with manure. Significantly, the lowest values of these parameters were recorded on the unfertilized treatment (0). These findings are consistent with those from Barneix [68] and Hlisnikovský and Kunzová [54].

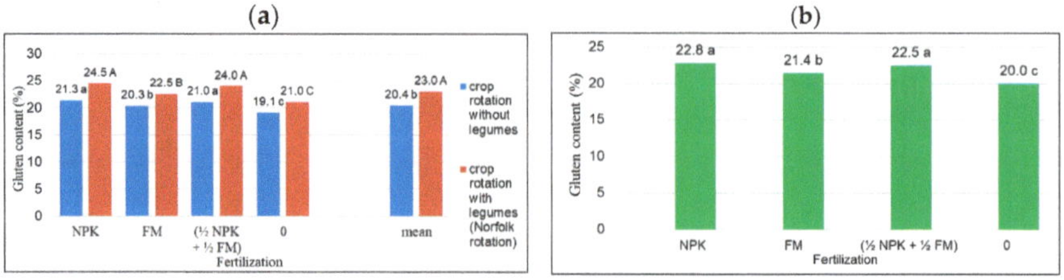

Figure 7. Gluten content in winter wheat grain depending on fertilization in two rotations (**a**) and only fertilization (**b**), means of years 2011, 2015, and 2019. Individual letters indicate homogeneous groups of means, mean values marked with the same letters do not differ significantly at $\alpha = 0.05$, different letters indicate significant differences at $\alpha = 0.05$. In subfigure (**a**), lowercase letters refer to crop rotation without legumes; capital letters refer to crop rotation with legumes (Norfolk rotation).

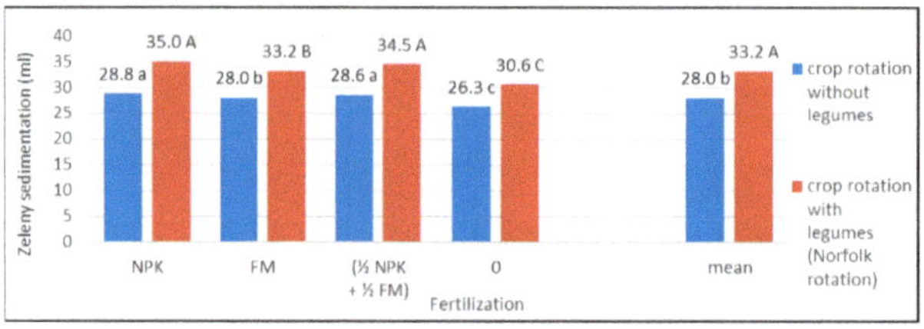

Figure 8. Zeleny sedimentation in winter wheat grain depending on rotation and fertilization (means of years 2009, 2013, and 2021). Individual letters indicate homogeneous groups of means, mean values marked with the same letters do not differ significantly at $\alpha = 0.05$, different letters indicate significant differences at $\alpha = 0.05$; lowercase letters refer to crop rotation without legumes; capital letters refer to crop rotation with legumes (Norfolk rotation).

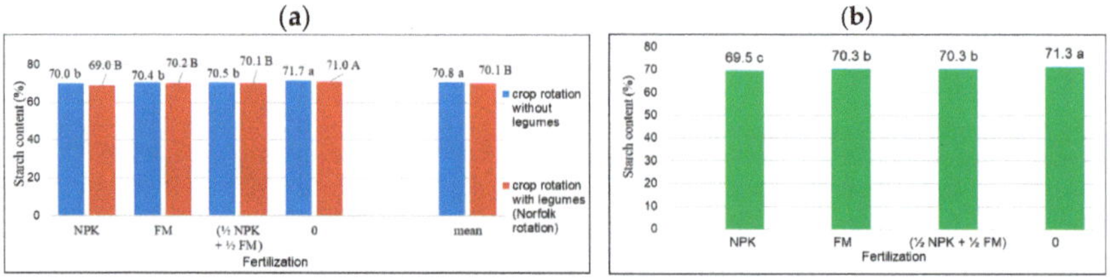

Figure 9. Starch content in winter wheat grain depending on fertilization in two rotations (**a**) and only fertilization (**b**), means of years 2011, 2015, and 2019. Individual letters indicate homogeneous groups of means, mean values marked with the same letters do not differ significantly at $\alpha = 0.05$, different letters indicate significant differences at $\alpha = 0.05$. In subfigure (**a**), lowercase letters refer to crop rotation without legumes; capital letters refer to crop rotation with legumes (Norfolk rotation).

The cultivation of red clover in the rotation also had a positive effect on the protein content, gluten content, and Zeleny sedimentation index in winter wheat grains.

The results in Figure 9 show that the lowest starch content was found in wheat grains fertilized only with minerals, and the highest in unfertilized wheat grains. Red clover as a pre-crop for winter wheat also resulted in a significant reduction in the starch content in the grain. Hlisnikovský and Kunzová [54] reported similar results and found a significantly lower starch content in wheat grain fertilized with mineral and organic fertilizers compared to the control.

The evaluation of the general effect of the experimental factors and years, as well as their interaction results of the ANOVA (p-values) for grain yield, which is the most important variable, are presented in Table 6. In the case of interaction with year, the only significant interaction was year x fertilization for grain yield of spring barley. It proves that the effect of fertilization on the grain yield of barley was modified by weather conditions in various years. It is probably because of the higher sensitivity of spring crops on water stress in drought seasons and the effect of fertilization, which modifies the effect of water stress [69].

Table 6. Results of ANOVA (*p*-values), which present the main effects and interactions for the grain yield of spring barley and winter wheat.

Effect	Spring Barley	Winter Wheat
Year	<0.001	0.008
Crop rotation	0.229	0.007
Fertilization	<0.001	0.001
Year × crop rotation	0.232	0.680
Year × fertilization	0.014	0.079
Crop rotation × fertilization	0.100	0.144

4. Conclusions

Our research, based on many years of static experiments on black soil, has shown that the Norfolk rotation with red clover as well as varied fertilization and weather conditions in the years of research affect the yield of cereal plants.

1. The highest grain yields of spring barley (5.7 t ha^{-1}) were ensured by mineral fertilization (NPK) and mineral fertilization in combination with manure ($\frac{1}{2}$NPK + $\frac{1}{2}$FM). Significantly lower yields were found in the area fertilized only with manure (FM) (5.0 t ha^{-1}), and the lowest were found (2.3 t ha^{-1}) in the absence of fertilization since 1955. However, the highest yields of winter wheat grain were recorded in the treatments with exclusive mineral fertilization (NPK) (6.4 t ha^{-1}). Significantly lower yields were found in the treatments where mineral fertilizers were used in combination with manure ($\frac{1}{2}$NPK + $\frac{1}{2}$FM) (5.7 t ha^{-1}) and only manure (FM) (5.1 t ha^{-1}), and the lowest yields were found in the absence of fertilization since 1955.
2. The use of undersown red clover in cultivation did not significantly affect the yield of spring barley grain, while clover as a forecrop for winter wheat created favorable conditions for plant growth. This is evidenced by significantly higher grain yields of winter wheat grown after red clover compared to the yields of this plant obtained after winter rapeseed (on average 5.7 t ha^{-1} vs. 4.8 t ha^{-1}, respectively).
3. Mineral (NPK) and mineral fertilization with manure ($\frac{1}{2}$NPK + $\frac{1}{2}$FM) and the cultivation of red clover in the rotation had a beneficial effect on the quality of spring barley and winter wheat grain. Mineral fertilization and mineral fertilization with manure resulted in an increase in the content of protein, wet gluten, and the Zeleny sedimentation index in winter wheat grain.

Author Contributions: Conceptualization, I.S.; formal analysis, I.S. and D.G.; investigation, I.S., K.P.-Z., and R.L.; methodology, I.S.; supervision, I.S.; visualization, D.G.; writing—original draft, I.S., K.P.-Z., D.G., and R.L.; writing—review and editing, I.S. and D.G. All authors have read and agreed to the published version of the manuscript.

Funding: This research received no external funding.

Institutional Review Board Statement: Not applicable.

Data Availability Statement: The data presented in this study are available upon request from the corresponding author. The data are not publicly available due to the fact that the data come from long-term experiments that are still being conducted. We will not provide the full data set, as we will use them in scientific articles prepared in the future.

Conflicts of Interest: Authors declare no conflicts of interest.

References

1. Bailey, K.L.; Gossen, B.D.; Lafond, G.P.; Watson, P.R.; Derksen, D.A. Effect of Tillage and Crop Rotation on Root and Foliar Diseases of Wheat and Pea in Saskatchewan from 1991 to 1998: Univariate and Multivariate Analyses. *Can. J. Plant Sci.* **2001**, *81*, 789–803. [CrossRef]
2. Sieling, K.; Christen, O. Crop Rotation Effects on Yield of Oilseed Rape, Wheat and Barley and Residual Effects on the Subsequent Wheat. *Arch. Agron. Soil Sci.* **2015**, *61*, 1531–1549. [CrossRef]
3. Barbieri, P.; Pellerin, S.; Nesme, T. Comparing Crop Rotations between Organic and Conventional Farming. *Sci. Rep.* **2017**, *7*, 13761. [CrossRef] [PubMed]
4. Jalli, M.; Huusela, E.; Jalli, H.; Kauppi, K.; Niemi, M.; Himanen, S.; Jauhiainen, L. Effects of Crop Rotation on Spring Wheat Yield and Pest Occurrence in Different Tillage Systems: A Multi-Year Experiment in Finnish Growing Conditions. *Front. Sustain. Food Syst.* **2021**, *5*, 647335. [CrossRef]
5. Kuś, J.; Jończyk, K. Wpływ Międzyplonów i Sposobu Uprawy Roli Na Plonowanie Roślin i Zawartość Azotu w Glebie. *Rocz. Nauk. Rol. Ser. A* **2000**, *114*, 83–95.
6. Podolska, G. Wartość Technologiczna Ziarna Pszenicy Ozimej w Zależności Od Dawki Nawożenia Azotem. *Przegląd Zbożowo-Młyn.* **2003**, *47*, 12–14.
7. Suwara, I.; Lenart, S.; Gawrońska-Kulesza, A. Wzrost i Plonowanie Pszenicy Ozimej Po 50 Latach Zróżnicowanego Nawożenia i Zmianowania. *Acta Agrophys.* **2007**, *10*, 695–704.
8. Niewiadomski, W.; Zawiślak, K. Tolerancja Jęczmienia Jarego Na Uproszczenie Zmianowania. *Zesz. Probl. Postępów Nauk. Rol.* **1979**, *218*, 31–37.
9. Smagacz, J. Porównanie Plonowania Jęczmienia Jarego i Pszenżyta Jarego Uprawianych Po Przedplonach Zbożowych. *Pamiętnik Puławski* **1998**, *112*, 193–200.
10. Wesołowski, M.; Kwiatkowski, C. Reakcja Nie Których Odmian Jęczmienia Jarego Na Uprawę w Krótkotrwałej Monokulturze. *Fragm. Agron.* **1997**, *14*, 36–42.
11. Woźniak, A. Wpływ Wsiewek Poplonowych i Nawożenia Organicznego Na Plonowanie, Zachwaszczenie i Zdrowotność Pszenżyta Ozimego w Monokulturze. Czesc I. Plon Ziarna. *Zesz. Probl. Postępów Nauk. Rol.* **2000**, *470*, 75–82.
12. Czuba, R. Mikroelementy We Współczesnych Systemach Nawożenia. *Zesz. Probl. Post. Nauk. Roln.* **2000**, *471*, 161–169.
13. Mazur, T.; Sądej, W. Działanie Wieloletniego Nawożenia Obornikiem, Gnojowicą i Nawozami Mineralnymi Na Plon Roślin i Białka. *Zesz. Probl. Postępów Nauk. Rol.* **1999**, *465*, 181–194.
14. Mazur, T.; Sądej, W. Działanie Wieloletniego Nawożenia Organicznego i Mineralnego Na Plon Jęczmienia Jarego i Pszenicy Ozimej Uprawianych Na Glebie Lekkiej. *Zesz. Probl. Postępów Nauk. Rol.* **2002**, *484*, 377–384.
15. Gaweł, E. Rola Roślin Motylkowatych Drobnonasiennych w Gospodarstwie Rolnym. *Woda-Śr.-Obsz. Wiej.* **2011**, *11*, 73–91.
16. Alaru, M.; Talgre, L.; Luik, A.; Tein, B.; Eremeev, V.; Loit, E. Barley Undersown with Red Clover in Organic and Conventional Systems: Nitrogen Aftereffect on Legume Growth. *Zemdirb.-Agric.* **2017**, *104*, 131–138. [CrossRef]
17. Sadowski, A. *Zrównoważony Rozwój Gospodarstw Rolnych z Uwzględnieniem Wpływu Wspólnej Polityki Rolnej Unii Europejskiej (Sustainable Development of Agricultural Holdings Taking into Account the Impact of the European Union's Common Agricultural Policy)*; Rozprawy Naukowe 447; Wydawnictwo Uniwersytetu Przyrodniczego w Poznaniu: Poznań, Poland, 2012.
18. Czyżewski, A. Teoriopoznawcze Przesłanki Rozwoju Rolnictwa Rodzinnego (Theoretical Prerequisites for the Development of Family Farming). In *Ekonomiczne Mechanizmy Wspierania i Ochrony Rolnictwa Rodzinnego w Polsce i Innych Państwach Unii Europejskiej*; Chlebicka, A., Ed.; FAPA: Warszawa, Poland, 2015; pp. 9–30.
19. Santín-Montanyá, M.I.; Zambrana, E.; Fernández-Getino, A.P.; Tenorio, J.L. Dry Pea (*Pisum sativum* L.) Yielding and Weed Infestation Response, under Different Tillage Conditions. *Crop Prot.* **2014**, *65*, 122–128. [CrossRef]
20. Martin-Guay, M.-O.; Paquette, A.; Dupras, J.; Rivest, D. The New Green Revolution: Sustainable Intensification of Agriculture by Intercropping. *Sci. Total Environ.* **2018**, *615*, 767–772. [CrossRef]
21. Horn, R.; Taubner, H.; Wuttke, M.; Baumgartl, T. Soil Physical Properties Related to Soil Structure. *Soil Tillage Res.* **1994**, *30*, 187–216. [CrossRef]
22. Suwara, I.; Gawrońska-Kulesza, A.; Korc, M. Wpływ Systemów Nawożenia Na Kształtowanie Się Wybranych Właściwości Fizycznych Gleby Lekkiej. *Fragm. Agron.* **2005**, *1*, 290–297.
23. Suwara, I. *Rola Wieloletniego Nawożenia w Kształtowaniu Wybranych Właściwości Gleby Lekkiej Ze Szczególnym Uwzględnieniem Stosunków Wodno–Powietrznych*; Wydawnictwo SGGW: Warszawa, Poland, 2010.
24. Lopushniak, V. Influence of Fertilizing Schemes in the Crop Rotation System on the Organic Matter and Nitrogen Content in the Dark-Grey Podzolized Soil in the Western Forest-Steppe of the Ukraine. *Pol. J. Soil Sci.* **2011**, *44*, 19–24.
25. Edmeades, D.C. The Long-Term Effects of Manures and Fertilizers on Soil Productivity and Quality: A Review. *Nutr. Cycl. Agroecosyst.* **2003**, *66*, 165–180. [CrossRef]
26. Twarog, S. Organic Agriculture. A Trade and Sustainable Development Opportunity for Developing Countries. In *Trade and Environment Review 2006*; United Nations: New York, NY, USA; Geneva, Switzerland, 2006; pp. 141–223.
27. Karlen, D.L.; Stott, D.E. A Framework for Evaluating Physical and Chemical Indicators of Soil Quality. In *SSSA Special Publications*; Doran, J.W., Coleman, D.C., Bezdicek, D.F., Stewart, B.A., Eds.; Soil Science Society of America and American Society of Agronomy: Madison, WI, USA, 1994; pp. 53–72. ISBN 978-0-89118-930-5.

28. Zhao, N.; Ma, J.; Wu, L.; Li, X.; Xu, H.; Zhang, J.; Wang, X.; Wang, Y.; Bai, L.; Wang, Z. Effect of Organic Manure on Crop Yield, Soil Properties, and Economic Benefit in Wheat-Maize-Sunflower Rotation System, Hetao Irrigation District. *Plants* **2024**, *13*, 2250. [CrossRef] [PubMed]
29. Zeweld, W.; Van Huylenbroeck, G.; Girmay, T.; Speelman, S. Impacts of Social and Psychological Issues on Adoption Behaviour for Agroforestry Systems, Crop Rotation and Compost Fertiliser in the Northern Ethiopia. In Proceedings of the 2017 International Congress, Parma, Italy, 28 August–1 September 2017. [CrossRef]
30. Gaudin, A.; Westra, S.; Loucks, C.; Janovicek, K.; Martin, R.; Deen, W. Improving Resilience of Northern Field Crop Systems Using Inter-Seeded Red Clover: A Review. *Agronomy* **2013**, *3*, 148–180. [CrossRef]
31. Suwara, I.; Gawrońska-Kulesza, A. Yielding and Field Structure of Winter Wheat after Different Forecrops in Depend on Rate of Nitrogen Fertilization. *Fragm. Agron.* **1995**, *2*, 216–217.
32. Suwara, I.; Gawrońska-Kulesza, A. Rola Przedplonu w Ograniczeniu Nawożenia Azotem Pod Pszenicę Ozimą. *Zesz. Probl. Postępów Nauk. Rol.* **1997**, 2011–2014.
33. Knudsen, T.M.; Hauggaard-Nielsen, H.; Jørnsgård, B.; Steen Jensen, E. Comparison of Interspecific Competition and N Use in Pea–Barley, Faba Bean–Barley and Lupin–Barley Intercrops Grown at Two Temperate Locations. *J. Agric. Sci.* **2004**, *142*, 617–627. [CrossRef]
34. Song, Y.N.; Zhang, F.S.; Marschner, P.; Fan, F.L.; Gao, H.M.; Bao, X.G.; Sun, J.H.; Li, L. Effect of Intercropping on Crop Yield and Chemical and Microbiological Properties in Rhizosphere of Wheat (*Triticum aestivu*m L.), Maize (*Zea mays* L.), and Faba Bean (*Vicia faba* L.). *Biol. Fertil. Soils* **2007**, *43*, 565–574. [CrossRef]
35. Peoples, M.B.; Craswell, E.T. Biological Nitrogen Fixation: Investments, Expectations and Actual Contributions to Agriculture. *Plant Soil* **1992**, *141*, 13–39. [CrossRef]
36. Peoples, M.B.; Herridge, D.F.; Ladha, J.K. Biological Nitrogen Fixation: An Efficient Source of Nitrogen for Sustainable Agricultural Production? *Plant Soil* **1995**, *174*, 3–28. [CrossRef]
37. Peoples, M.B. Legumes Root Nitrogen in Cropping System Nitrogen Cycling. *Graine Legume* **2001**, *33*, 8–9.
38. Chinthapalli, B. A Comparative Study on the Effect of Organic and Inorganic Fertilizers on Agronomic Performance of Faba Bean (*Vicia faba* L.) and Pea (*Pisum sativum* L.). *Agric. For. Fish.* **2015**, *4*, 263. [CrossRef]
39. Seleiman, M.F.; Ibrahim, M.E.; Darwish, I.H.; Hardan, A.N.M. Effect of Mineral and Organic Fertilizers on Yield and Quality of Some Egyptian and Omani Wheat Cultivars. *Menoufia J. Plant Prod.* **2021**, *6*, 351–372. [CrossRef]
40. Korschens, M. The Importance of Long-Term Field Experiments for Soil Science and Environmental Research—A Review. *Plant Soil. Environ.* **2006**, *69*, 113–125.
41. Merbach, W.; Deubel, A. Long-Term Field Experiments—Museum Relics or Scientific Challenge? *Plant Soil Environ.* **2008**, *54*, 219–226. [CrossRef]
42. Kunzová, E.; Hejcman, M. Yield Development of Winter Wheat over 50 Years of FYM, N, P and K Fertilizer Application on Black Earth Soil in the Czech Republic. *Field Crops Res.* **2009**, *111*, 226–234. [CrossRef]
43. Hejcman, M.; Kunzová, E. Sustainability of Winter Wheat Production on Sandy-Loamy Cambisol in the Czech Republic: Results from a Long-Term Fertilizer and Crop Rotation Experiment. *Field Crops Res.* **2010**, *115*, 191–199. [CrossRef]
44. Kabała, C.; Charzyński, P.; Chodorowski, J.; Drewnik, M.; Glina, B.; Greinert, A.; Hulisz, P.; Jankowski, M.; Jonczak, J.; Łabaz, B.; et al. Polish Soil Classification, 6th Edition—Principles, Classification Scheme and Correlations. *Soil Sci. Annu.* **2019**, *70*, 71–97. [CrossRef]
45. Carmer, S.G.; Walker, W.M. Pairwise Multiple Comparisons of Treatment Means in Agronomic Research. *J. Agron. Educ.* **1985**, *14*, 19–26. [CrossRef]
46. TIBCO Software Inc. *Statistica (Data Analysis Software System)*; Version 13; TIBCO Software Inc.: Palo Alto, CA, USA, 2017.
47. Stumpe, H.; Wittenmayer, L.; Merbach, W. Effects and Residual Effects of Straw, Farmyard Manuring, and Mineral-N Fertilization at Field F of the Long-Term Trial in Halle (Saale), Germany. *J. Plant Nutr. Soil Sci.* **2000**, *163*, 649–656. [CrossRef]
48. Ellmer, F.; Baumecker, M. Static Nutrient Depletion Experiment Thyrow. Results after 65 Experimental Years. *Arch. Agron. Soil Sci.* **2005**, *51*, 151–161. [CrossRef]
49. Merbach, W.; Herbst, F.; Eißner, H.; Schmidt, L.; Deubel, A. Influence of Different Long-Term Mineral–Organic Fertilization on Yield, Nutrient Balance and Soil C and N Contents of a Sandy Loess (Haplic Phaeozem) in Middle Germany. *Arch. Agron. Soil Sci.* **2013**, *59*, 1059–1071. [CrossRef]
50. Hlisnikovský, L.; Zemanová, V.; Roman, M.; Menšík, L.; Kunzová, E. Long-Term Study of the Effects of Environment, Variety, and Fertilisation on Yield and Stability of Spring Barley Grain. *Plants* **2024**, *13*, 2745. [CrossRef] [PubMed]
51. Wanic, M.; Treder, K.; Myśliwiec, M.; Brzezin, G.M. Wpływ Wsiewek Międzyplonowych Na Cechy Biometryczne i Plonowanie Jęczmienia Jarego. *Fragm Agron.* **2012**, *29*, 160–171.
52. Andruszczak, S.; Kraska, P.; Kwiecińska-Poppe, E.; Pałys, E. Wpływ Wsiewek Międzyplonowych Oraz Stosowania Herbicydu Chwastox Extra 300 SL Na Plon Ziarna i Elementy Plonowania Jęczmienia Jarego Uprawianego w Monokulturze. *Biul. Inst. Hod. I Aklim. Roślin* **2011**, *259*, 147–156. [CrossRef]
53. Jiang, D.; Hengsdijk, H.; Dai, T.-B.; De Boer, W.; Jing, Q.; Cao, W.-X. Long-Term Effects of Manure and Inorganic Fertilizers on Yield and Soil Fertility for a Winter Wheat-Maize System in Jiangsu, China. *Pedosphere* **2006**, *16*, 25–32. [CrossRef]
54. Hlisnikovský, L.; Kunzová, E. Effect of Mineral and Organic Fertilizers on Yield and Technological Parameters of Winter Wheat (*Triticum aestivum* L.) on Illimerized Luvisol. *Pol. J. Agron.* **2014**, *17*, 18–24.

55. Blecharczyk, A.; Zawada, D.; Sawinska, Z.; Małecka-Jankowiak, I.; Waniorek, W. Wpływ Następstwa Roślin i Nawożenia Na Plonowanie Pszenicy Ozimej. *Fragm. Agron.* **2019**, *36*, 27–35.
56. Ailincăi, C.; Ailincăi, D.; Zbant, M.; Mercuş, A.; Ţopa, D.; Cara, M. The Effect of Different Fertilization Systems on Wheat Yield, Erosion and Fertility of Eroded Soils from the Moldavian Plateau. *Cercet. Agron. În Mold.* **2007**, *4*, 5–14.
57. Barzegar, A.R.; Yousefi, A.; Daryashenas, A. The Effect of Addition of Different Amounts and Types of Organic Materials on Soil Physical Properties and Yield of Wheat. *Plant Soil* **2002**, *247*, 295–301. [CrossRef]
58. Berzsenyi, Z.; Győrffy, B.; Lap, D. Effect of Crop Rotation and Fertilisation on Maize and Wheat Yields and Yield Stability in a Long-Term Experiment. *Eur. J. Agron.* **2000**, *13*, 225–244. [CrossRef]
59. Norwood, C.A. Dryland Winter Wheat as Affected by Previous Crops. *Agron. J.* **2000**, *92*, 121–127. [CrossRef]
60. Blecharczyk, A.; Śpitalniak, J.; Małecka, I. Wpływ Doboru Przedplonów Oraz Systemów Uprawy Roli i Nawożenia Azotem Na Plonowanie Pszenicy Ozimej. *Fragm. Agron.* **2006**, *23*, 273–286.
61. Buczek, J.; Bobrecka-Jamro, D.; Szpunar-Krok, E.; Tobiasz-Salach, R. Plonowanie Pszenicy Ozimej w Zależności Od Przedplonu i Stosowanych Herbicydów. *Fragm. Agron.* **2009**, *26*, 7–14.
62. Smagacz, J.; Kuś, J. Wpływ Długotrwałego Stosowania Płodozmianów Zbożowych Na Plonowanie Zbóż. *Fragm. Agron.* **2010**, *27*, 119–134.
63. Amato, G.; Ruisi, P.; Frenda, A.S.; Di Miceli, G.; Saia, S.; Plaia, A.; Giambalvo, D. Long-Term Tillage and Crop Sequence Effects on Wheat Grain Yield and Quality. *Agron. J.* **2013**, *105*, 1317–1327. [CrossRef]
64. Małecka Jankowiak, I.; Blecharczyk, A.; Sawinska, Z.; Waniorek, W. Wpływ następczy łubinów i grochu na plonowanie pszenicy ozimej w zależności od uprawy roli i nawożenia azotem. *Fragm. Agron.* **2018**, *35*, 67–79. [CrossRef]
65. Subedi, M.; Ghimire, B.; Bagwell, J.W.; Buck, J.W.; Mergoum, M. Wheat End-Use Quality: State of Art, Genetics, Genomics-Assisted Improvement, Future Challenges, and Opportunities. *Front. Genet.* **2023**, *13*, 1032601. [CrossRef]
66. Assefa, A.; Girmay, G.; Alemayehu, T.; Lakew, A. Performance Evaluation and Stability Analysis of Malt Barley (*Hordeum vulgare* L.) Varieties for Yield and Quality Traits in Eastern Amhara, Ethiopia. *CABI Agric. Biosci.* **2021**, *2*, 31. [CrossRef]
67. Meng, G.; Rasmussen, S.K.; Christensen, C.S.L.; Fan, W.; Torp, A.M. Molecular Breeding of Barley for Quality Traits and Resilience to Climate Change. *Front. Genet.* **2023**, *13*, 1039996. [CrossRef]
68. Barneix, A.J. Physiology and Biochemistry of Source-Regulated Protein Accumulation in the Wheat Grain. *J. Plant Physiol.* **2007**, *164*, 581–590. [CrossRef] [PubMed]
69. Gan, Y.T.; Lafond, G.P.; May, W.E. Grain Yield and Water Use: Relative Performance of Winter vs. Spring Cereals in East-Central Saskatchewan. *Can. J. Plant Sci.* **2000**, *80*, 533–541. [CrossRef]

agriculture

MDPI

Article

Monitoring and Signaling of the Most Important Aphid Species in the Territory of Greater Poland and Silesia Provinces

Kamila Roik [1,*], Anna Tratwal [1], Sandra Małas [1] and Jan Bocianowski [2]

1 Department of Monitoring and Signaling of Agrophages, Institute of Plant Protection-National Research Institute, Władysława Węgorka 20, 60-318 Poznan, Poland; a.tratwal@iorpib.poznan.pl (A.T.); s.malas@iorpib.poznan.pl (S.M.)
2 Department of Mathematical and Statistical Methods, Poznań University of Life Sciences, Wojska Polskiego 28, 60-637 Poznan, Poland; jan.bocianowski@up.poznan.pl
* Correspondence: k.roik@iorpib.poznan.pl

Abstract: Aphids are significant pests affecting crop yields both through direct feeding and as vectors of viruses. The monitoring focused on 10 of the most important aphid species. This study investigates the dynamics of aphid populations in two Polish regions, Winna Góra (Greater Poland Province) and Sośnicowice (Silesia Province), over a five-year period (2019–2023) using Johnson suction traps. Data collection covered species composition, migration timing, and seasonal variations in aphid abundance. Dominance patterns were assessed using a species-specific index, and inter-regional comparisons were analyzed through correlation and principal component analysis. Results indicate notable population peaks during autumn, suggesting this period is optimal for implementing control measures. The Johnson traps proved valuable for timely pest monitoring, offering predictive potential for future aphid migration, particularly in relation to virus-transmitting species critical to plants.

Keywords: aphids; flight dynamics; monitoring; Johnson's suction trap

1. Introduction

During the growing season, crop plants are exposed to attack by many species of agrophages, such as pathogenic microorganisms, weeds, and pests [1]. Aphids are among the most important pests of crop plants, which can cause significant yield losses, both directly, as a result of feeding on plants, and indirectly, when they are vectors of viruses [2–7]. Indirect damages usually cause greater yield losses than direct pests.

Aphids are highly adaptive to changing environmental conditions [8–10]. The evolutionary adaptation of aphids to respond quickly to temperature changes is primarily due to their small body size and rapid development of generations, which consequently allows their populations to grow rapidly [11–13]. According to the study, 764 taxa (species and subspecies), occurring in 167 genera, had been recorded in Poland by 2015. About 100–150 species of aphids are economically important pests worldwide [5,14,15].

The threat to crops from aphids, which occurs every year, makes it necessary to conduct systematic and long-term monitoring of the dynamics of their flights. Such studies, which are a form of detailed recording of 24 h aphid flights, make it possible to track what structural changes are taking place in the afidofauna of the studied area. Knowledge of the spread of aphids and their colonization of new areas is a key element in forecasting their emergence and signaling. The spread of aphids and their colonization of new areas, which can travel considerable distances, was the subject of research conducted using a Johnson suction trap. The Johnson suction trap offers a unique advantage over other aphid sampling methods due to its continuous, autonomous operation and capacity to sample at heights. Unlike pan and sticky traps, which are passive and tend to capture only aphids actively flying close to the ground, the Johnson trap can provide data on aphid migration and dispersal at higher elevations [16].

The aim of the ongoing research was to analyze the species composition, flight timing, and abundance dynamics of the most important aphid species.

2. Material and Methods

Winna Góra and Sośnicowice are two locations where Johnson suction traps have been installed, each with distinct environmental characteristics. The trap in Winna Góra is situated on the grounds of the Experimental Field Station, a branch of the Institute of Plant Protection–National Research Institute (IPP-NRI), in western Poland (Greater Poland Province, Środa Wielkopolska County, coordinates: latitude 52.20548, longitude 17.44712). The surrounding area comprises agricultural fields (mainly winter wheat, winter barley, winter rape, and sugar beet), forests, and small towns. Similarly, in Sośnicowice, the trap is located within a branch of the IPP-NRI in southern Poland (10 km from Gliwice, Silesia Province, Gliwice County, coordinates: latitude 50.27099, longitude 18.54144) and is surrounded by fields, forests, and small towns. However, unlike Winna Góra, the Sośnicowice area also includes water reservoirs. These regional and environmental differences may impact local microclimatic conditions and the diversity of insect species present, which is relevant for studying the populations and migration patterns of insects such as aphids. The steel structure of the aspirator has an electric fan sucking air through a pipe 9 m long and 250 mm in diameter, and the total height of the aspirator is 12.2 m. This is the optimal height for studying aphid flights from remote areas. The samples collected represent the status of migratory fauna in an area with a radius of up to 80 km from the aspirator site [17–19]. The device operated from the beginning of May, at the time of the beginning of the migration of the first aphids until the end of October, i.e., until the cessation of autumn flights, from 6:00 a.m. to 10:00 p.m. Samples with trapped insects were taken every day at a fixed time (12:00 p.m.). Aphids were selected from among the collected insects, which were then determined into species and counted based on available keys and catalogs for identifying aphid species [20–24]. The labeled and identified material was preserved in 70% propanol. The individual dominance index for a grouping of migratory aphids caught with an aspirator was calculated according to the formula [25]:

$$D = n/N \cdot 100 \text{ [in \%]},$$

where n is the number of individuals of a given species present in the sample at a certain time and N is the number of all individuals of aphids caught with an aspirator at a certain time.

Five classes of dominance were adopted: D5—eudominants—more than 10% of the number of individuals of each species in the sample; D4—dominants—5.1–10.0%; D3—subdominants—2.1–5.0%; D2—recedents—1.1–2.0%; and D1—subrecedents—less than 1.0%.

An analysis of variance was conducted to evaluate the effect of the differentiating factors studied (years and locations) and year-by-location interaction on aphid abundance. The interdependence of the abundance of individual aphid species in the studied environments (combinations of localities and survey years) was assessed using Pearson's linear correlation and Spearman's rank correlation. The similarity in the grouping of aphid species based on data observed in all environments (combinations of localities and survey years) taken together was examined multidimensionally by applying principal component analysis. All statistical analyses were performed using the Genstat 23.1 package (VSN International Genstat for Windows 23rd Edition. VSN International, Hemel Hempstead, UK, 2023).

3. Results and Discussion

Systematic trapping of aphids allowed analysis of changes in the abundance and species composition of winged aphids in 2019–2023. The 10 economically important aphid species trapped in the study were *Acyrthosiphon pisum*, *Aphis fabae*, *Aphis frangulae*, *Aphis nasturtii*, *Anoecia corni*, *Brevicoryne brassicae*, *Metopolophium dirhodum*, *Myzus persicae*, *Rhopalosiphum padi*, and *Sitobion avenae* [8]. In the two localities where aspirator trapping was conducted, different, though sometimes very similar, dates of the first aphid flights were observed in each year of the study. The most similar was in 2020, where the difference in the first flights of all species aphids between Winna Gora and Sosnicowice was a maximum of 4 days (Tables 1 and 2). The intensity of the flights of these species varied depending on the season and the year of observation. The total number of aphids caught using the Johnson aspirator in all years of the study was 165,253 in both localities (Winna Gora—73,526 and Sosnicowice—91,727) (Tables 3 and 4). The highest number of aphids was found in 2022 in both localities, i.e., 42,860 in Winna Góra and 55,455 in Sosnicowice. Much less was caught in 2020 (Winna Góra—9411, Sosnicowice—11,769) and 2021 (Winna Góra—9315, Sosnicowice—13,258), while less numerous in 2019 Winna Góra—5049, Sosnicowice—5057) and 2023 (Winna Góra—6892, Sosnicowice—6188). The first flights of aphids take place at the turn of May and June.

Table 1. Dates of the beginning of first flights economically 10 important species of aphids caught by Johnson suction trap in Winna Góra.

	Year/Beginning of First Flights				
Species	**2019**	**2020**	**2021**	**2022**	**2023**
Rhopalosiphum padi L.	10 May	9 May	24 May	8 May	12 May
Anoecia corni F.	12 May	13 May	5 June	6 June	22 May
Sitobion avenae F.	19 May	12 May	21 May	10 May	19 May
Metopolophium dirhodum Walk.	3 June	8 June	16 June	19 May	22 May
Myzus persicae Sulz.	9 May	11 May	7 June	14 May	2 June
Brevicorynae brassicae L.	16 May	14 May	11 June	5 June	23 June
Aphis fabae Scop.	11 May	11 May	10 May	8 June	12 May
Aphis frangulae Kalt and *Aphis nasturtii* Kalt.	8 June	16 June	16 June	7 May	2 June
Acyrthosiphon pisum Harris	11 May	16 May	1 June	16 May	18 May

Table 2. Dates of the beginning of first flights economically 10 important species of aphids caught by Johnson suction trap in Sośnicowice.

	Year/Beginning of First Flights				
Species	**2019**	**2020**	**2021**	**2022**	**2023**
Rhopalosiphum padi L.	2 May	8 May	19 May	11 May	12 May
Anoecia corni F.	25 May	9 May	13 June	6 June	20 May
Sitobion avenae F.	17 May	12 May	25 May	12 May	8 May
Metopolophium dirhodum Walk.	5 June	12 June	9 June	25 May	21 May
Myzus persicae Sultz	7 May	9 May	30 June	15 May	27 May
Brevicorynae brassicae L.	11 May	12 May	3 June	2 June	16 June
Aphis fabae Scop.	19 May	8 May	17 May	7 June	11 May
Aphis frangulae Kalt and *Aphis nasturtii* Kalt.	-	12 June	8 June	4 June	3 June
Acyrthosiphon pisum Harris	5 May	20 May	3 June	8 June	11 May

Table 3. Species composition of winged aphid morphs in the years of studies in Winna Góra.

Species	Year/Number of Aphids										Total	%	Class of Dominance
	2019	%	**2020**	%	**2021**	%	**2022**	%	**2023**	%			
Rhopalosiphum padi L.	2472	48.96	5766	61.27	6498	69.76	36,452	85.04	5598	81.22	56,786	77.23	D5
Anoecia corni F.	819	16.22	1604	17.04	1411	15.15	2634	6.15	445	6.46	6913	9.40	D4
Sitobion avenae F.	501	9.92	472	5.01	401	4.3	347	0.81	58	0.84	1778	2.42	D3
Metopolophium dirhodum Walk.	139	2.75	116	1.23	51	0.55	62	0.14	137	1.99	505	0.69	D1
Myzus persicae Sulz.	381	7.54	625	6.64	325	3.49	1326	3.1	403	5.85	3060	4.16	D3
Brevicorynae brassicae L.	63	1.25	71	0.75	66	0.7	70	0.16	20	0.29	290	0.40	D1
Aphis fabae Scop.	427	8.46	488	5.19	383	4.11	1588	3.7	125	1.81	3011	4.10	D3
Aphis frangulae Kalt and *Aphis nasturtii* Kalt.	10	0.2	8	0.09	3	0.03	4	0.01	75	1.09	100	0.13	D1
Acyrthosiphon pisum Harris	237	4.70	261	2.77	177	1.9	377	0.88	31	0.45	1083	1.47	D2
Total	5049	100	9411	100	9315	100	42,860	100	6892	100	**73,526**	100	-

Table 4. Species composition of winged aphid morphs in the years of studies in Sośnicowice.

Species	Year/Number of Aphids										Total	%	Class of Dominance
	2019	%	**2020**	%	**2021**	%	**2022**	%	**2023**	%			
Rhopalosiphum padi L.	3364	66.52	6954	59.1	9159	69.08	49,120	88.57	3855	62.3	72,452	78.99	D5
Anoecia corni F.	1344	26.58	3830	32.54	3344	25.22	5458	9.84	1882	30.41	15,858	17.29	D5
Sitobion avenae F.	56	1.1	132	1.12	148	1.11	58	0.10	54	0.87	448	0.49	D1
Metopolophium dirhodum Walk.	52	1.03	46	0.39	118	0.9	9	0.01	44	0.71	269	0.29	D1
Myzus persicae Sulz.	178	3.52	490	4.16	221	1.67	380	0.69	196	3.17	1465	1.6	D2
Brevicorynae brassicae L.	7	0.13	22	0.2	35	0.26	128	0.23	41	0.66	233	0.25	D1
Aphis fabae Scop.	12	0.24	228	1.88	145	1.1	269	0.49	75	1.21	729	0.79	D1
Aphis frangulae Kalt and *Aphis nasturtii* Kalt.	0	0	6	0.05	14	0.1	6	0.01	12	0.19	38	0.04	D1
Acyrthosiphon pisum Harris	44	0.87	61	0.56	74	0.56	27	0.05	29	0.47	235	0.26	D1
Total	5057	100	11,769	100	13,258	100	55,455	100	6188	100	**91,727**	100	-

The localities, as well as their specificities (the surrounding effects, the weather conditions), all have an impact on the development of the aphids and on the difference in their abundance. In 2019, the average annual temperature across Poland was 10.2 °C, 2.4 °C higher than the 1971–2000 norm. Spring was warm in most parts of Poland, including Greater Poland, and very warm in areas of Silesia. Summer, as well as autumn throughout Poland, was extremely warm. In terms of precipitation, 2019 was classified as normal. The average air temperature in 2020 in Poland was 9.9 °C. Particularly warm months were February and August, while very cool months included May. The average precipitation in 2020 in Poland was 645.4 mm. The driest month was April. The area's average air temperature in 2021 was 8.7 °C in Poland. The year was classified as a thermally normal year. The amount of precipitation in 2021 in Poland was 627.4 mm and was classified as a normal years. The area's average air temperature in 2022 was 9.5 °C in Poland, and it was a very warm and dry year (average precipitation in Poland was 534.4 mm). The area's average air temperature in 2023 was 10.0 °C in Poland, as much as 1.3 °C higher than the multi-year average from 1991–2020; 2023 was an extremely warm year with precipitation levels at average levels. The most abundant species in all years of the study, both in Winna Gora and Sosnicowice, was *R. padi*, the percentage of which ranged from 49 to 88.6% in individual years. Also in the studies of other scientists, this species was also caught in the greatest numbers [26,27]. In the study, aphids belonging to the species *A. corni* were also very numerous, the percentage of which ranged from 6.1 to 32.5% in individual years. Aphids of the species *M. persicae*, the proportion of which ranged from 0.7 to 7.5%, and *A. fabae* (in Winna Gora from 1.8–8.5%) were also caught in greater numbers (in both localities).

The distribution of the observed aphid species in the pattern of the first two principal components was very efficient and explained a total of 99.95% of the total variation (Figure 1). The principal component analysis conducted allowed us to distinguish three groups of aphid species. The first group is the *R. padi* species; the second group is the *A. corni* species; and the third group is the other aphid species. The discriminant analysis performed shows a statistically significant effect of all combinations of localities and years on the first principal component. In contrast, the second principal component was not determined by any of the environments.

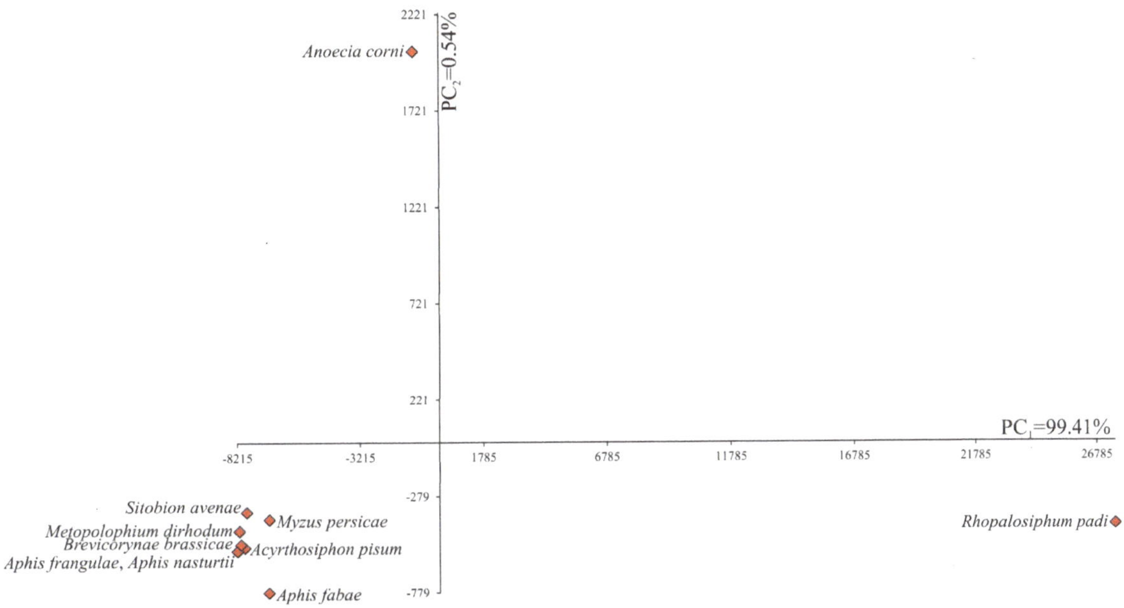

Figure 1. Distribution of the observed aphid species in the pattern of the first two principal components: PC_1 and PC_2.

Pearson's linear correlation coefficients of aphid abundance were statistically significant for all pairs of environments. They were all positive. For rank correlation, statistical significance was noted in most comparisons. No rank correlation of aphid abundance was observed in four cases: in Winna Góra between the years 2019–2023, 2021–2023, 2022–2023, and between Sośnicowice 2022 and Winna Góra 2023 (Table 5).

Based on the methodology of Zlotkowski and Bandyk [28], an index of individual dominance was determined, where the values of multi-year average abundance totals of individual insect species were taken. Determination of dominance usually indicates the quantitative share of the studied species in specific ecosystems, for example, for individual water bodies or the grouping of different organisms in a specific place within a geographical region. The study of species structure in relation to the grouping of migratory aphids is a new form of analysis of the course of their migration. In Winna Góra, the most numerous group consisted of species belonging to the subdominants, two species were classified as subrecedents, and one species each was classified in the eudominant classes. On the other hand, in Sośnicowice, the most numerous were the subrecedents grouping as many as 6 aphid species, two species were classified into the eudominant group, and one species was classified into the recedents. The percentage of *R. padi* in all years of the study in both localities exceeded 10%, which classified this species into the eudominants, and the abundance of *A. corni* species in Winna Góra at the turn of 5 years classified this species in the dominant group and in Sośnicowice in the eudominant group (Tables 3 and 4).

Table 5. Pearson's linear correlation coefficients (below the diagonal) and Spearman's rank correlation coefficients (above the diagonal) of aphid abundance between environments.

Location		Winna Góra					Sośnicowice				
	Year	2019	2020	2021	2022	2023	2019	2020	2021	2022	2023
Winna Góra	2019	1	0.933 ***	0.983 ***	0.883 **	0.633	0.850 **	0.933 ***	0.933 ***	0.800 **	0.883 **
	2020	0.993 ***	1	0.917 ***	0.950 ***	0.767 *	0.883 **	1.000 ***	0.967 ***	0.900 ***	0.950 ***
	2021	0.987 ***	0.998 ***	1	0.900 ***	0.55	0.800 **	0.917 ***	0.900 ***	0.850 **	0.867 **
	2022	0.963 ***	0.981 ***	0.990 ***	1	0.633	0.733 *	0.950 ***	0.850 **	0.917 ***	0.867 **
	2023	0.96 ***	0.980 ***	0.988 ***	0.999 ***	1	0.767 *	0.767 *	0.800 **	0.633	0.817 **
Sośnicowice	2019	0.971 ***	0.986 ***	0.979 ***	0.945 ***	0.947 ***	1	0.883 **	0.950 ***	0.733 *	0.867 **
	2020	0.948 ***	0.959 ***	0.945 ***	0.894 ***	0.894 ***	0.990 ***	1	0.967 ***	0.900 ***	0.950 ***
	2021	0.975 ***	0.989 ***	0.985 ***	0.956 ***	0.955 ***	0.999 ***	0.985 ***	1	0.850 **	0.967 ***
	2022	0.967 ***	0.986 ***	0.994 ***	0.998 ***	0.997 ***	0.961 ***	0.914 ***	0.970 ***	1	0.917 ***
	2023	0.958 ***	0.971 ***	0.961 ***	0.917 ***	0.917 ***	0.997 ***	0.998 ***	0.993 ***	0.936 ***	1

* $p < 0.05$; ** $p < 0.01$; *** $p < 0.001$.

Most of the trapped aphid species are dioecious species that re-migrate to secondary hosts in autumn. The dynamics of aphid flights during the years of the study indicate a significant disparity in aphid abundance in autumn compared to spring and summer. May and June were taken as the spring months, July and August as the summer months, and September and October as the autumn months. In studies by other researchers, the seasonal dynamics of aphid flight abundance were similar [26,29–32], and sometimes a higher proportion of aphids was recorded during spring migration [33,34]. This is due to the occurrence of increasingly long and warm autumns, which allow aphids to develop longer (Figures 2 and 3).

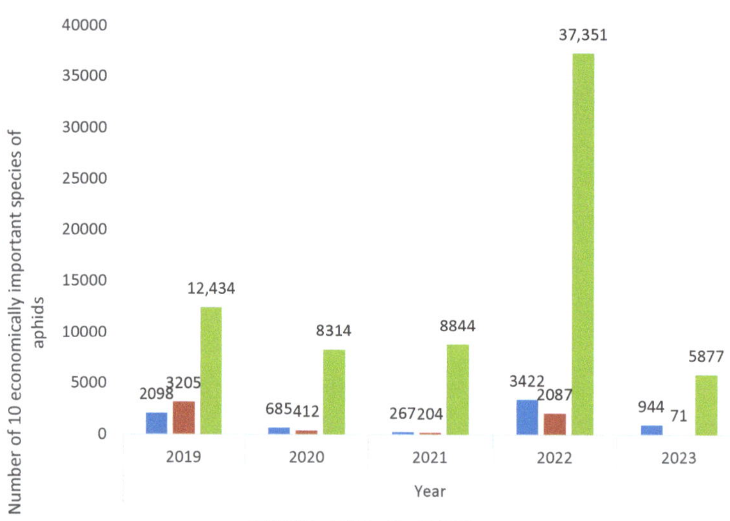

Figure 2. Comparison of the number of 10 important species caught during the spring, summer, and autumn of 2019–2023 in Winna Góra.

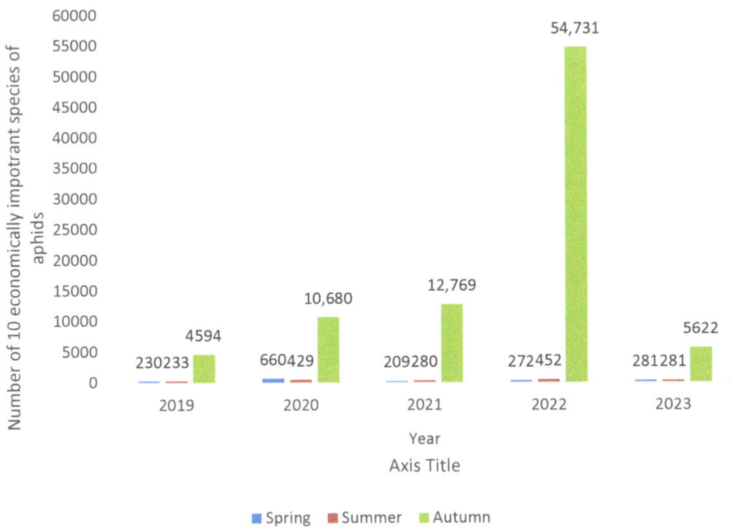

Figure 3. Comparison of the number of 10 important species caught during the spring, summer, and autumn in 2019–2023 in Sośnicowice.

Daily aphid trapping makes it possible to track the intensity and timing of the flight of individual species, which is particularly important for signaling threats, mainly from aphid vectors of viruses. Remigrations of various species recorded with the aspirator can provide a basis for forecasting the intensity of their appearance in the following year. In the case of short-term forecasting, the aspirator is an extremely important tool for ascertaining the presence of aphids in the air shortly before they colonize crops, which allows rapid decision-making on protective treatment [35–40]. The guidelines of integrated control, which apply to all professional users of crop protection products, clearly indicate the priority of using preventive methods crop of protection.

One of the most important elements of integrated control is prevention. Using the results obtained from aphid trapping with the Johnson aspirator is a good example of such measures. First of all, it is important that the presence of specific aphid species in the trapped material indicates the threat of these species in crop fields in about 10 to 14 days. Secondly, the results obtained from the trapping are representative of the area within a radius of about 80 km from the device [35,37,38,40].

The results obtained are very useful for planning plantation protection strategies, especially in the context of the observed climate change, i.e., warming. The increased number of days with warmer weather in autumn has a significant impact on increasing the threat from pests, including aphids. The use of equipment such as the Johnson aspirator is an important element in supporting compliance with the principles of integrated protection.

4. Conclusions

In five years of aphid trapping with the Johnson suction traps, there was a clear dominance of two species: *R. padi* and *A. corni* in relation to the total number of aphids caught.

Based on the abundance of aphids trapped in 2019–2023, there was a clear dominance of autumn migrations of these insects compared to flights observed in spring and summer at both locations.

The use of the Johnson suction traps in the autumn period is very useful in determining the optimal timing of chemical treatment against *R. padi*, the main vector of barley yellow dwarf virus (BYDV). Accuracy and precision in determining the correct timing of aphid control affect reducing the level of chemization, which is the main goal of integrated protection.

Author Contributions: Conceptualization, K.R., A.T. and S.M.; methodology, K.R., A.T. and J.B.; software, K.R. and J.B.; validation, A.T. and J.B.; formal analysis, K.R. and S.M.; investigation, K.R. and S.M.; resources, A.T. and J.B.; data curation, K.R.; writing—original draft preparation, K.R., A.T., S.M. and J.B.; writing—review and editing, K.R., A.T., S.M. and J.B.; visualization, S.M. and J.B.; supervision, A.T. and J.B.; project administration, K.R.; funding acquisition, K.R. All authors have read and agreed to the published version of the manuscript.

Funding: This research received no external funding.

Institutional Review Board Statement: Not applicable.

Data Availability Statement: The data presented in this study are available on request from the corresponding author.

Conflicts of Interest: The authors declare no conflict of interest.

References

1. Mrówczyński, M.; Walczak, F.; Korbas, M.; Paradowski, A.; Roth, M. Zmiany klimatyczne a zagrożenia roślin rolniczych przez agrofagi. *Stud. I Rap. IUNG–PIB* **2009**, *17*, 139–148.
2. Ruszkowska, M. Przekształcenia cyklicznej partenogenezy mszycy *Rhopalosiphum padi* (L.) (Homoptera, Aphidoidea)—znaczenie zjawiska w adaptacji środowiskowej. *Rozpr. Nauk. Inst. Ochr. Roślin* **2002**, *8*, 63.
3. Singh, R.; Singh, G. Aphids. In *Polyphagous Pests of Crops*; Springer: Singapore, 2021. [CrossRef]
4. Fingu-Mabola, J.C.; Francis, F. Aphid–Plant–Phytovirus Pathosystems: Influencing Factors from Vector Behaviour to Virus Spread. *Agriculture* **2021**, *11*, 502. [CrossRef]
5. Ruszkowska, M. Uwarunkowania klimatyczne w rozprzestrzenianiu najważniejszych wektorów chorób wirusowych na zbożach w badanych regionach Polski. [Permanent and cyclic parthenogenesis of *Rhopalosiphum padi* (L.) (Homoptera: Aphidoidea) across different climate regions in Poland]. *Prog. Plant Prot. Postępy W Ochr. Roślin* **2006**, *46*, 276–283.
6. Strażyński, P. *Population Structure of Rhopalosiphum padi (Linnaeus, 1758) /Hemiptera, Aphidoidea/ in Wielkopolska Region in 2003–2008 in the Context of Winter Cereals Threat of BYDV Expansion. s. 91–105. W: Aphids and Other Hemipterous Insects;* Goszczyński, W., Herczek, A., Leszczyński, B., Łabanowski, G., Podsiadło, E., Rakauskas, R., Ruszkowska, M., Wilkaniec, B., Wojciechowski, W., Eds.; The John Paul II Catholic University of Lublin: Lublin, Poland, 2010; Volume 16, p. 125.
7. Borodynko-Filas, N.; Pruszyński, G.; Strażyński, P. Wirus żółtaczki rzepy (Turnip Yellows Virus, TuYV) i jego wektory—nowe zagrożenie w uprawie rzepaku. *Streszcz. 57. Ses. Nauk. IOR—PIB Poznań 9–10 Lutego* **2017**, *2017*, 72.
8. Strażyński, P.; Ruszkowska, M.; Węgorek, P. Dynamika lotów mszyc w latach 2008–2010 najliczniej odławianych w Poznaniu aspiratorem Johnsona. [Flights dynamics of aphids caught numerously by Johnson's suction trap in Poznań in 2008–2010]. *Prog. Plant Prot. Postępy W Ochr. Roślin* **2011**, *51*, 213–216.
9. Csorba, A.B.; Dinescu, S.; Pircalabioru, G.G.; Fora, C.G.; Bálint, J.; Loxdale, H.D.; Balog, A. Aphid adaptation in a changing environment through their bacterial endosymbionts: An overview, including a new major cereal pest (*Rhopalosiphum maidis* (Fitch) scenario. *Symbiosis* **2024**, *93*, 139–152. [CrossRef]
10. Parry, H.R. Cereal aphid movement: General principles and simulation modelling. *Mov. Ecol.* **2013**, *1*, 14. [CrossRef]
11. Carter, W. *Owady a Choroby Roślin*; Państwowe Wydawnictwo Rolnicze i Leśne: Warszawa, Poland, 1971; p. 494.
12. Leszczyński, B.; Jóźwik, B.; Laskowska, I.; Szynkarczyk, S. Wiosenne migracje mszycy czeremchowo-zbożowej (*Rhopalosiphum padi*) na pszenżyto. *Prog. Plant Prot.* **1998**, *38*, 289–291.
13. Wang, X.J.; Ma, C.S. Can laboratory-reared aphid populations reflect the thermal performance of field populations in studies on pest science and climate change biology? *J. Pest Sci.* **2023**, *96*, 509–522. [CrossRef]
14. Wojciechowski, W.; Depa, Ł.; Kanturski, M.; Wegierek, P.; Wieczorek, K. Annotated checklist of aphids (Hemiptera: Aphidomorpha) of Poland. *Pol. J. Entomol.* **2015**, *84*, 384–420. [CrossRef]
15. Blackman, R.L.; Eastop, V.E. Aphids on the World Plants. An Online Identification and Information Guide. 2022. Available online: http://www.aphidsonworldsplants.info/ (accessed on 9 September 2024).
16. Heatcote, G.D. The comparison of yellow cylindrical, flat and water traps, and of Johnson suction traps, for sampling aphids. *Ann. App. Biol.* **1957**, *45*, 133–139. [CrossRef]
17. Taylor, L.R. An improved suction trap for insect. *Ann. Appl. Biol.* **1951**, *38*, 582–591. [CrossRef]
18. Allison, D.; Pike, K.S. An inexpensive suction trap and its use in an aphid monitoring network. *J. Agric. Entomol.* **1988**, *5*, 103–107.
19. Złotkowski, J. Zmienność nasilenia pojawu mszycy kapuścianej (*Brevicoryne brassicae* L.) rejestrowana za pomocą nowej metody odławiania w latach 1973-75. *Pr. Nauk. Inst. Ochr. Roślin* **1997**, *XIX*, 167–176.
20. Szelegiewicz, H. Katalog Fauny Polski. In *Część XXI. Zeszyt 4. Mszyce—Aphidodea*; PWN: Warszawa, Poland, 1968; p. 316.
21. Petrović, A. Sizing the Knowledge Gap in Taxonomy: The Last Dozen Years of Aphidiinae Research. *Insects* **2022**, *13*, 170. [CrossRef]
22. Müller, F.P. *Mszyce–Szkodniki Roślin: Terenowy Klucz do Oznaczania*; Instytut Zoologii Polskiej Akademii Nauk.: Warszawa, Poland, 1976; p. 119.

23. Ward, S.E.; Umina, P.A.; Macfadyen, S.; Hoffmann, A.A. Hymenopteran Parasitoids of Aphid Pests within Australian Grain Production Landscapes. *Insects* **2021**, *12*, 44. [CrossRef]
24. Taylor, L.R. *A Handbook for Aphid Identification*; Euraphid–Rothamsted Experimental Station: Rothamsted, UK, 1980; p. 171.
25. Kasprzak, K.; Niedbała, W. *Wskaźniki Biocenotyczne Stosowane Przy Porządkowaniu i Analizie Danych w Badaniach ilościowych. s. 397–416. W: Metody Stosowane w Zoologii Gleby*; Górny, M., Grüm, L., Eds.; Państwowe Wydawnictwo Naukowe: Warszawa, Poland, 1981; p. 483.
26. Strażyński, P. Znaczenie rejestracji lotów ważnych gospodarczo gatunków i form mszyc w odłowach aspiratorem Johnsona w Poznaniu w latach 2003–2005 w integrowanych metodach ochrony roślin. [Importance of registration of flights of economically important species and forms of aphids in catches by Johnson's suction trap in Poznań in 2003–2005 in integrated plant protection methods]. *Prog. Plant Prot. Postępy W Ochr. Roślin* **2006**, *46*, 395–398.
27. Gałuszka, A.; Drzewiecki, S.; Wolski, A. Dynamika lotów ważniejszych gospodarczo gatunków mszyc odławianych w latach 2007-2011 przy użyciu aspiratora Johnsona na terenie województwa śląskiego. *Prog. Plant Prot.* **2015**, *55*, 216–220.
28. Złotkowski, J.; Bandyk, A. Charakterystyka zmian w strukturze składu gatunkowego mszyc migrujących w odłowach aspiratorem Johnsona w latach 1999–2010. [Characteristic of the changes in the structure of aphid species composition based on catches of Johnson's suction trap in 1999–2010]. *Prog. Plant Prot. Postępy W Ochr. Roślin* **2012**, *52*, 252–258. [CrossRef]
29. Strażyński, P. *Dynamics of Aphid Seasonal Flights in Johnson's Suction Trap in Poznań in 2003–2004. s. 175–183. W: Aphids and Other Hemipterous Insects*; Cichocka, E., Goszczyński, W., Ruszkowska, M., Wilkaniec, B., Nowak, P., Kowalski, G., Eds.; Agricultural University of Poznań: Poznań, Poland, 2005; Volume 11, p. 212.
30. Mahas, J.W.; Mahas, J.B.; Ray, C.; Kesheimer, A.; Steury, T.D.; Conzemius, S.R.; Crow, W.; Gore, J.; Greene, J.K.; Kennedy, G.G.; et al. The Spatiotemporal Distribution, Abundance, and Seasonal Dynamics of Cotton-Infesting Aphids in the Southern U.S. *Insects* **2023**, *14*, 639. [CrossRef] [PubMed]
31. Luquet, M.; Poggi, S.; Buchard, C.; Plantegenest, M.; Tricault, Y. Predicting the seasonal flight activity of Myzus persicae, the main aphid vector of Virus Yellows in sugar beet. *Pest Manag. Sci.* **2023**, *79*, 4508–4520. [CrossRef] [PubMed]
32. Złotkowski, J. Sezonowe zmiany w dynamice migracji mszyc w okolicach Winnej Góry (Wielkopolska) w latach 2007–2008. [Seasonal changes in aphid migration dynamics in the surrounding area of Winna Góra (Wielkopolska district) in 2007–2008]. *Prog. Plant Prot. Postępy W Ochr. Roślin* **2009**, *49*, 1242–1246.
33. Złotkowski, J. Rejestracja sezonowych lotów ważnych gospodarczo mszyc w odłowach aspiratorem zainstalowanym na terenie Pracowni Doświadczalnictwa Polowego w Winnej Górze w latach 2001–2004. [Registration of seasonal flights of economically important aphids in catches by suction trap installed in the Field Experimental Station in Winna Góra in the years 2001–2004]. *Prog. Plant Prot. Postępy W Ochr. Roślin* **2005**, *45*, 1229–1232.
34. Slavíková, L.; Fryč, D.; Kundu, J.K. Analysis of Twenty Years of Suction Trap Data on the Flight Activity of Myzus persicae and Brevicoryne brassicae, Two Main Vectors of Oilseed Rape Infection Viruses. *Agronomy* **2024**, *14*, 1931. [CrossRef]
35. Ruszkowska, M.; Strażyński, P. Monitoring lotów ważnych gospodarczo gatunków mszyc w niektórych rejonach Polski, Czech i Niemiec. [Monitoring of flights of economically important species of aphids in some regions of Poland, the Czech Republic and Germany]. *Prog. Plant Prot. Postępy W Ochr. Roślin* **2004**, *44*, 1114–1117.
36. Lau, D.; Sampaio, M.V.; Salvadori, J.R.; da Silva Pereira, P.R.V.; dos Santos, C.D.R.; Engel, E.; Panizzi, A.R.; Júnior, A.L.M. Historical and Contemporary Perspectives on the Biological Control of Aphids on Winter Cereals by Parasitoids in South America. *Neotrop. Entomol.* **2023**, *52*, 172–188. [CrossRef]
37. Ruszkowska, M.; Strażyński, P. *Mszyce na Oziminach*; Instytut Ochrony Roślin: Poznań, Poland, 2007; 23 ss.
38. Ruszkowska, M.; Strażyński, P. Profilaktyka w ochronie zbóż przed chorobą żółtej karłowatości jęczmienia. [Prophylaxis in the protection of cereals against barley yellow dwarf viruses]. *Prog. Plant Prot. Postępy W Ochr. Roślin* **2007**, *47*, 363–366.
39. Eigenbrode, S.D.; Adhikari, S. Climate change and managing insect pests and beneficials in agricultural systems. *Agron. J.* **2023**, *115*, 2194–2215. [CrossRef]
40. Złotkowski, J. Znaczenie pułapki ssącej Johnsona w badaniach migracji mszyc w Polsce w latach 1973–2010. [The importance of Johnson's suction trap in the study of aphid migration in Poland in the years 1973–2010]. *Prog. Plant Prot. Postępy W Ochr. Roślin* **2011**, *51*, 158–163.

www.ingramcontent.com/pod-product-compliance
Lightning Source LLC
Chambersburg PA
CBHW042010260926
51850CB00020B/2963
* 9 7 8 3 7 2 5 8 4 1 8 9 9 *